Table of Contents

Acknowledgments

The author wishes to express his appreciation to Brian Jirout, a doctoral student at the Georgia Institute of Technology; Marcus Jackson, an undergraduate student at Xavier University; and Lauren Binger, an undergraduate student at Smith College, for assistance in collecting information for this project. The author also wishes to thank NASA's Emerging Space Office, which provided a grant to pursue this research.

Executive Summary and Findings

The study that follows investigates and analyzes historical episodes in America in which the federal government undertook public-private efforts to complete critical activities valued for their public good. This combination largely resulted from a lack of either sufficient political will to fund them entirely out of the public treasury or insufficient profit motive for private firms to undertake them for purely business reasons. The six case studies include the following: 1) the development of the transcontinental railroad, supported by a unique land-grant approach to subsidy; 2) support for the airline industry through legislation, appropriate regulation, and subsidies to grow a robust air transport capability; 3) the regulatory regime put into place with the rise of the telephone industry and the creation of a government-sponsored monopoly that eventually had to be broken up; 4) government sponsorship of Antarctic scientific stations that evolved into a public-private partnership (PPP) over time; 5) the fostering of a range of public works projects and their success or failure over time; and 6) the establishment of scenic and cultural conservation zones in the United States and ways to balance economic development with preservation.

With the rise of a range of private-sector entrepreneurial firms interested in pursuing space commerce, the process whereby their efforts might be incubated, fostered, and expanded comes to the fore as an important public policy concern in a way never before present in the Space Age. In the United States, and really nowhere else in the world, we are witnessing the convergence of several powerful economic forces. These include the need to restore American capability to reach low-Earth orbit (LEO) for the servicing of the International Space Station (ISS), the rise of a hospitality/tourism/entertainment industry interested in space, the development of expansive remote sensing and other applications in Earth orbit, and the possibilities envisioned for opening commercial space activities in the cislunar region.

Through these case studies, we explore how to apply more effectively already-tested models of government support for commercial activities, as well as the interactions of both the public and private spheres in a new opportunity zone in space. In each case, a summation yields a range of key points. The following paragraphs relate key conclusions.

Transcontinental Railroad: The approach taken by government involvement in 19th-century transcontinental railroad development remains valid to some degree for orbital space operations. The government offered the following six inducements for private development:

1. Land grants as a means of offering potential future revenue, tied to success in creating the railroad system.
2. Direct government appropriations to the company involved in the endeavor.
3. Waivers/modifications to taxes and other regulatory requirements.
4. Contracts for services once capability was demonstrated.
5. Government endorsement and backing of corporate bonds/assets.
6. Indirect support for related but supplemental elements of the railroad transportation system.

In every case, these government initiatives were intended to leverage (and not replace) existing private funding, especially additional industry and venture capital.

To those six, we might add the following:

- Private financing supplemented with government loans.
- Property and patent rights granted to participating firms.
- Broadly construed revenues produced from transportation and other fees.

Regardless, one must ask these critical questions in the context of developing new space transportation structures: "How important, in the final analysis, is cheaper access to space? Is it really the key to the future growth of space activities?" This seems to be at the cusp of what will go into any stimulation of private space transportation effort.

Commercial Air Transportation: Between 1915 and the 1970s, government officials in the United States undertook a series of critical initiatives designed to create a commercial airline industry in private hands. Washington lawmakers saw the necessity of fostering new technology for the purposes of national security, economic competitiveness, and pride and prestige. That last reason was in no small measure because although Americans had invented the airplane in 1903, by 1914 leadership in the technology had moved to Europe—the United States had been left in the dust. Catching up became an important driver for federal investment. Government organizations took a multifaceted approach: military investment, research and development, regulatory efforts aimed at both promoting safety and efficiency and expanding operations, and direct subsidies to commercial entities until the 1960s. Congress could have established a national airline run by civil servants, but instead created a favorable climate for private investment in airlines. For instance, the U.S. Congress established the National Advisory Committee for Aeronautics (NACA) in 1915 to conduct research on flight, and in 1921 New York and New Jersey created a port authority with the power to issue bonds and collect fees for airfields.

In terms of space transportation, there are several lessons to be drawn from the aviation experience. Like the NACA, government agencies could conduct basic research and transfer that knowledge to private firms. In addition, the National Aeronautics and Space Administration (NASA) could transfer its operational responsibility to private carriers. Congress could also create the authority—modeled on various earlier efforts such as the Overseas Private Investment Corporation—to provide loans/insurance to space line firms. Either the U.S. government or states could establish spaceport authorities to manage operations from the ground to orbit; federal agencies could also regulate routes and fares. Many of these efforts are already under way, and we are on the verge of seeing a new age of entrepreneurial space transportation efforts. There are, however, challenges to this approach, not the least of which is that NASA has a critical path with specific milestone deadlines and is hesitant to change this approach; the loans/insurance incentives may not produce services in time; and liability issues are especially burdensome. Nonetheless, major steps have been taken toward this capability in the last decade.

Telecommunications: Following the invention of the telephone in 1876, the federal government could have owned and operated telephone service—it did so during World War I—or it could have allowed a totally open market. Instead, it established phone companies as regulated monopolies under the Federal

Communications Commission (FCC), with monopolistic privileges left in place until 1980. In essence, the following structure emerged:

- Regulatory agencies provided patents, granted monopoly status, and chartered corporations.
- The U.S. Attorney General allowed the American Telephone and Telegraph Company (AT&T) to control telephone service as a regulated monopoly (1913).
- AT&T established Bell Laboratories (1925); Bell Labs developed the first orbiting communications satellite (Telstar 1, 1962).
- Congress created the Communication Satellite Corporation (COMSAT), a public-private corporation with monopoly status, to promote satellite communications (1962).
- COMSAT represented the United States in the formation of INTELSAT and became its managing company.

Might the U.S. government foster a private space communications system that could serve the needs of all users on a commercial basis rather than having NASA own its own Tracking and Data Relay Satellite System (TDRSS) satellites? What is the future of space communications? Will the government encourage private entrepreneurs to construct, own, operate, and use lunar communications networks, Mars communications networks, deep space networks? A major challenge: recent experience (Iridium, Global Positioning System [GPS]) suggests that the cost of establishing certain space communications networks exceeds likely revenues.

Antarctica: Antarctica has a legal status similar to that of the Moon. It is utilized primarily for scientific research, and no nation can claim its land. Yet basic supplies and logistic support for U.S. operations on the continent are provided by nonfederal organizations. Might this become a possibility in the future on the ISS or the Moon? Fostering such an approach to space activities could mean that control of orbital lunar assets would remain with NASA, which would select and fund science projects, oversee policy, and cycle personnel as necessary. The operation of these stations, however, could fall to a company with experience in remote locations, staffed by its own employees. Transportation to and from these stations could also be provided by outside organizations. At the same time, commercial activities could be encouraged.

Public Works: Frequently in the history of the United States, the federal government has developed critical infrastructure, often for its national security purposes, but quickly leading to economic development. At times, it has relied wholly upon private entrepreneurs. One of the most creative approaches to this process has been the use of the government or quasi-government development commissions to develop resources as a public good. There are many instances of this approach to public-private partnership. For example, in the General Mining Act of 1872, the U.S. government set up an uncontrolled but highly entrepreneurial structure that emphasized the principle that discovery conferred ownership. It left a legacy of riches and ruin that few wanted to repeat. More recently, commissions have been formed to create a more controlled development of the resource. Examples include the Isthmian Canal Commission (1904); the Bonneville Power Administration (1937); and the subject of this discussion, the Tennessee Valley Authority (TVA, 1933). This entity was decentralized, not a conventional government agency. Congress provided at least initial appropriation, but the organization was intended to become self-sustaining while delivering a public service. Corporate entities associated with it were empowered to borrow and spend as well as market goods and services. TVA served an important economic and social purpose and, in the process, served as the catalyst for the wholesale transformation of the region.

In the context of lunar development, might an organization similar to TVA be capable of commercially developing the Moon? Questions abound:

- Should it begin with the establishment of a lunar development commission/corporation?
- Would a commission/corporation start by building and managing lunar infrastructure for NASA?
- Would this be followed by an effort to spur economic development?

National Parks: In terms of applicability to the space frontier, the experience of the National Park Service is most germane in terms of space tourism efforts. When Congress created the U.S. National Park Service in 1916 to conserve natural and historical resources "by such means as will leave them unimpaired," a key component was to assist the public in reaching those scenic wonders. Accordingly, park managers, recognizing the need for public support to encourage future preservation, allowed private entrepreneurs to commercialize the parks in such a manner as to encourage public visitation. Accordingly, they encouraged railroad companies and other concessionaires to build hotels and related facilities in the national parks. Those concessionaires then paid fees, which the National Park Service used to build additional roads and trails. Americans rode and drove to the national parks, vastly expanding tourism and creating the family vacation tradition.

Government officials, as well as policy, could likewise encourage private-sector development in space tourism, both in low-Earth orbit and on the Moon. The following possibilities exist:

- Public officials could expand the use of government facilities by private entrepreneurs as a means of encouraging public use and visitation.
- Private firms could pay fees, which government agencies could then use to expand and develop facilities.
- Government could create a favorable regulatory climate for space tourism.
- Private citizens could then experience space through both remote access and direct participation.

Beyond these very specific possibilities, NASA could also award lease contracts for habitation/support services of facilities in orbit and on the Moon. Baseline development and operational costs could then be funded by NASA lease. As an additional revenue stream, companies could then add tourism for marginal costs. Such an environment could create a public-private space ecology with efficiencies of operations achieved through economies of scale.

Introduction

With the rise of a range of private-sector entrepreneurial firms interested in pursuing space commerce in the first decade of the 21st century, the process whereby those efforts might be incubated, fostered, and expanded comes to the fore as an important public policy concern in ways never before confronted in the history of the Space Age.

The United States is witnessing the convergence of several powerful economic forces. These include the need to restore American capability to reach low-Earth orbit for the servicing of the International Space Station, the rise of a hospitality/tourism/entertainment industry interested in space, the place of remote sensing and other applications in Earth orbit as well as their commercial appeal, and the possibilities envisioned for opening commercial space activities in the cislunar region. Among those last efforts is the stimulation of several organizations to pursue a lunar lander and rover/hopper to capture a $25 million Google Lunar XPRIZE.

Because of this broad-based set of initiatives, there is a renewed necessity to explore historical analogs for the stimulation of private-sector investment in space activities. All of these represent various forms of public-private partnerships. The Council on Public-Private Partnerships defines this approach to big projects thusly: "A contractual agreement between a public agency (federal, state or local) and a private sector entity. Through this agreement, the skills and assets of each sector (public and private) are shared in delivering a service or facility for the use of the general public. In addition to the sharing of resources, each party shares in the risks and rewards potential in the delivery of the service and/or facility."[1]

There are many instances of these models of financing/governance/operations having been used throughout history. The National Council for Public-Private Partnerships offers these 10 statements about their broad use:

1. **Public-private partnerships are just what the name implies.** Public-private partnerships are a contractual arrangement whereby the resources, risks and rewards of both the public agency and private company are combined to provide greater efficiency, better access to capital, and improved compliance with a range of government regulations regarding the environment and workplace. The public's interests are fully assured through provisions in the contracts that provide for on-going monitoring and oversight of the operation of a service or development of a facility. In this way, everyone wins—the government entity, the private company and the general public.

2. **Public-private partnerships are more common than you may think.** Public-[p]rivate [p]artnerships have been in use in the United States for over 200 years and thousands are operating today. These contractual arrangements between government entities and private companies for the delivery of services or facilities [are] used for water/wastewater, transportation, urban development,

1 National Council for Public-Private Partnerships, *Testing Tradition: Assessing the Added Value of Public-Private Partnerships* (Washington, DC: National Council for Public-Private Partnerships, 2012), p. 2.

and delivery of social services, to name only a few areas of application. Today, the average American city works with private partners to perform 23 out of 65 basic municipal services. The use of partnerships is increasing because they provide an effective tool in meeting public needs, maintaining a high level of public control, [and] improving the quality of services, and [they] are more cost effective than traditional delivery methods.

3. **They are an essential tool in challenging economic times.** Even in the best of times, governments at all levels are challenged to keep pace with the demands of their constituencies. During periods of slow growth, government revenues are frequently not sufficient to meet spending demands, necessitating painful spending cuts or tax increases. Partnerships can provide a continued or improved level of service, at reduced costs. And equally important, partnerships can also provide the capital needed for construction of major facilities. By developing partnerships with private-sector entities, governments can maintain quality services despite budget limitations.

4. **Successful partnerships can lead to happy employees.** In many partnerships created today, public employees are retained and usually at equal or improved benefits. One of the greatest areas of improvement for employees is with opportunities for career growth—private companies spend two to three times more on training and personnel development than their public-sector counterparts, as a way of gaining the maximum efficiency out of every person, and the maximum amount of job satisfaction.

5. **Successful partnerships can lead to better public safety.** From Los Angeles to the District of Columbia, local governments have formed creative partnerships with private companies to enhance the safety of [their] streets and [their] citizens. By turning over the operation of parking meters or the processing of crime reports to private-sector partners, police officers can spend more time on the streets doing the jobs for which they are trained. This is particularly important as [h]ome[l]and [s]ecurity has risen as a concern for many.

6. **Partnerships give many children better educational opportunities.** In Virginia, public-private partnerships were instrumental in constructing over 30 new school buildings. By working with a private real estate development company, city and county school systems were able to build state-of-the-art facilities with a modern computer lab, gym and library. Often, allowing the private sector to utilize publicly-owned underutilized assets for commercial activities provides a major portion of the funding for these projects. Today, a number of other states are now following this example, driven by the need to address the problem of aging education infrastructures.

7. **Drivers appreciate public-private partnerships.** These are not easy times for America's roads and highways. Increasing numbers of vehicles means more roadway wear and tear and increasing traffic congestion. In states like California, Virginia and Texas, private-sector companies are working with state and local governments to build roads, making it possible to finance construction and upkeep without having to impose general tax increases. While tolling [is] one means of generating the revenue to cover the investment, in a number of cases

Transportation Oriented Development (TOD) of adjacent properties can provide a significant portion of the revenue stream.

8. **Clean, safe water is achieved through public-private partnerships.** The stringent health and environmental standards of the Safe Drinking Water Act and Clean Water Act have presented difficulties for some local governments without the budget flexibility to make major capital improvements in water and wastewater facilities. Public-private partnerships have enabled the construction of state-of-the-art water management facilities, while using efficient operations to hold down costs to ratepayers and provide a way of meeting those "un-funded mandates" from the federal government.

9. **Partnerships make the information revolution accessible to more Americans.** This is the age of information technologies, but there can be a hefty cost of getting a system operating. Through public-private partnerships, many governments are now able to fully participate in "E-government" with their constituents, or effectively coordinate government activities and budgets. Better service, improved tools and saving money are exactly what public-private partnerships are all about.

10. **Governments themselves are the biggest supporters of public-private partnerships**. While there can be substantial misperceptions about the value of partnerships, a look at who endorses them should clarify the picture. Federal agencies like the Environmental Protection Agency, the Department of Defense, and the Veterans Administration all use partnerships. And the number of state and local governments using this tool is even greater. For example, the U.S. Conference of Mayors is enthusiastically working with private-sector providers to discuss ways to make partnerships more effective. Numerous surveys indicate why—governments traditionally realize cost savings of 20 to 50 percent when the private-sector is involved in providing services.[2]

While there have long been public-private partnerships, the trends for the future are that they will gain in use over the next several years. It makes sense that these will become the norm for space activities as well. They are increasingly used in infrastructure and investment-heavy sectors as a means of leveraging public funding, operational efficiency, and risk.

There are many reasons for this development in the last third of the 20th century. The most significant has been the need to reduce costs, but of almost as much significance has been the lack of both political will and expertise in the public sector (see figure 1).[3] Regardless of this approach and its long use in American history, there are those who question whether or not public-private partnerships intrude on the democratic traditions of the United States. Critics have argued that these partnerships open up opportunities for corruption, for conspiring to provide poor-quality services, and for depriving

2 "Top Ten Facts About PPPS," National Council for Public-Private Partnerships, available online at *http://www.ncppp.org/ppp-basics/top-ten-facts-about-ppps/*, accessed 24 November 2013, 6:36 p.m.

3 National Council for Public-Private Partnerships, *For the Good of the People: Using Public-Private Partnerships To Meet America's Essential Needs* (Washington, DC: National Council for Public-Private Partnerships, 2012), p. 5.

underprivileged citizens of essential services. There is evidence to support these criticisms. There is also evidence to support counterarguments about success and win-win in public policy and the delivery of services to society.[4]

The reality is that public-private partnerships are neither a panacea for all ills of public policy nor a Pandora's box of troubles. They offer appropriate responses to problems of society, especially those that have technological, infrastructure, or enormous investment challenges. There are always tradeoffs.[5]

Should there be a desire to develop a public-private partnership to resolve some challenge there are a series of best practices that have been developed:

- *Best Practice #1: Understand the Uniqueness of the Project.* Over the past several decades PPPs have become an increasingly common tool throughout the world. This has resulted in experience and research which informed...
- *Best Practice #2: Include Everything, but Be Wary of Bias.* Not including every aspect of public costs can mean the difference between finding that a PPP is or is not cost-effective.
- *Best Practice #3: PCC [I]s Not a Final Step.* The PCC is just one way of exploring the possibility of public sector involvement and assessing [its] value for money. Furthermore, the PCC only provide[s] a partial picture. It is impossible to determine from only the public cost assessment whether a PPP will be beneficial without the comparative factor of what it would cost the private sector....
- *Best Practice #4: Be Realistic, Not Overly Optimistic.* Study after study has shown that public sector cost estimates are consistently too optimistic in terms of a project's cost, timeline, and ability to generate revenue. This has certainly proven to be the case for both PCCs and the costs of PPPs....
- *Best Practice #5: Identify Strengths AND Weaknesses*...deciding whether to provide a service, the public sector needs to conduct a complete calculation of the cost of the project....[6]

Although there are positives and negatives of public-private partnerships, when they are prepared carefully and managed with aplomb, they have proven useful in wide array of settings and industries.

We are accordingly seeing this in the spaceflight arena as well as in the development of more mundane utilities and infrastructure. As only one example, NASA's Kennedy Space Center in 2008 created its Center Planning and Development Office with the objective of pursuing public-private partnerships in

---

4 National Council for Public-Private Partnerships, *Critical Choices: The Debate Over Public-Private Partnerships and What It Means for America's Future* (Washington, DC: National Council for Public-Private Partnerships, 2003), p. 6.

5 Bernard L. Ungar, "Public-Private Partnerships: Factors To Consider When Deliberating Governmental Use as a Real Property Management Tool," General Accountability Office, October 1, 2001; Masafumi Hashimoto, "Public-Private Partnerships in Space Projects: An Analysis of Stakeholder Dynamics," M.A. Thesis, Massachusetts Institute of Technology (MIT), 2009.

6 Claire Goldbach, Valerie Goldman, Richard Phillips, and Ashlyn Seymour, "Public Cost Comparator for Public-Private Partnerships," American University Masters of Public Policy on behalf of the National Council for Private-Public Partnerships, 2011, pp. 23–25.

Figure 1

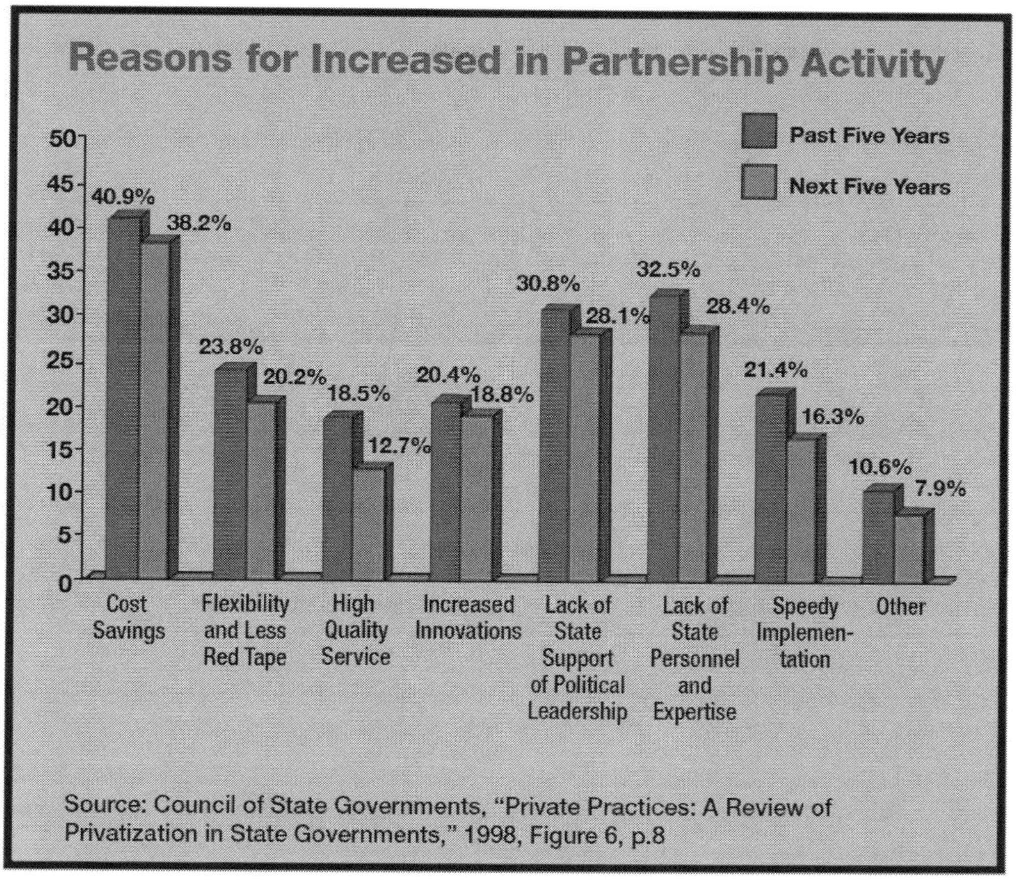

Source: Used with permission of the Council of State Governments.

the construction and management of launch capabilities at the nation's foremost spaceport.[7] There have been some notable successes (Telesat's Anik F2 and Hughes Early Bird), some spectacular failures (X-33/VentureStar and Galileo navigation satellite development), and some that have yet to prove themselves one way or another, especially NASA's present Commercial Crew Development (CCDev) program.[8]

In a pathbreaking 2013 study, "Public-Private Partnerships for Space Capability Development: Driving Economic Growth and NASA's Mission," the workability of these efforts has been studied for the spaceflight community in the following areas:

1. **Satellite Servicing:** On-orbit rescue, refueling, repositioning, repair, or inspection of satellites; can include mitigation and debris removal.
2. **Interplanetary Small Satellites:** Spacecraft that conduct missions beyond LEO and have a mass of <500 kg.

---

7 Amber N. Philman, "Public-Private Partnerships Expected To Launch a New Era of Space Flight at Kennedy Space Center," Florida League of Cities, August 2012, pp. 43–44.

8 Hossein Ghandeharian and Ali Shoamanesh, "Public Private Partnerships (PPP) for Development of the Space Sector," National Space Conference, Islamabad, Pakistan, 17–19 September 2012.

3. **Robotic Mining:** Extraction, processing, and transport of materials using autonomous or semi-autonomous equipment.
4. **Cargo Transportation Beyond LEO:** Transport of goods, equipment, and resources.
5. **Crew Transportation Beyond LEO:** Transport of crew, including astronauts and spaceflight participants.
6. **Microgravity Research for Biomedical Applications:** Used by researchers in biomedical science as a tool for discovery of medically-important applications.
7. **Liquid Rocket Engines:** Safe, reliable, cost effective engines.
8. **Wireless Power:** Transmitting power over both short and long distances without wires.
9. **Space Communications:** Technologies to transmit and relay data between satellites, vehicles in space, and assets on the ground.
10. **Earth Observation Data Visualization:** Comprises the massive volume and diversity of data collected by Earth observation satellite systems.[9]

Analysts found that public-private partnerships do help close the business case for U.S. companies, but only help—no more—because there is also an intrinsic role for the federal government in ensuring success in this arena, enticing private-sector investment in critical U.S. capabilities in space. In addition to other areas, they found that "[p]artnerships to develop commercial transportation systems capable of sending crew beyond LEO can provide NASA and other customers access to affordable crew transportation while stimulating new commercial markets." Furthermore, the study found, "[b]y leveraging **public-private partnerships** as a regular part of the agency's approach to **space capability development**, NASA's programs can be a fundamental driver of **U.S. economic growth in the 21st century**" (emphasis in original). This study offered the following recommendations:

1. Integrate economic analysis and market evaluation into Agency strategic decision-making and acquisition process for program formulation.
2. Engage in public-private partnerships in those areas that have attracted private capital, which have technical merit, and [that] contribute to achieving NASA's overall Mission and are in the national interest.
3. Develop a strategy and architecture for space exploration that includes public-private partnerships.[10]

In keeping with those recommendations, this study explores historical episodes in America in which the federal government fostered public-private partnerships to accomplish critical activities valued for their public good when they could not be achieved through commercial activities alone. These include the following six cases and their relationships to connected space activities: 1) the development of the transcontinental railroad, supported by a unique land-grant approach to subsidy; 2) support for the airline industry through legislation, appropriate regulation, and subsidies to grow a robust aerospace industry; 3) the regulatory regime put into place with the rise of the telephone industry and the creation

---

9 NASA Office of Strategy Formulation, "Public-Private Partnerships for Space Capability Development: Driving Economic Growth and NASA's Mission," 30 July 2013.

10 Ibid.

of a government-sponsored monopoly that eventually had to be broken up; 4) government sponsorship of Antarctic scientific stations that evolved into a public-private partnership over time; 5) the fostering of a range of public works projects and their success or failure over time; and 6) the establishment of scenic and cultural conservation zones in the United States and ways to balance economic development with preservation. Through these case studies, we may learn how to apply more effectively already-tested models of government support for commercial activities, as well as the interactions of both the public and private spheres in this new opportunity zone in space.

Specific questions in these six cases will revolve around the economic and political rationales for these activities—in greater specificity than I have seen before—and the application of those business and political models and practical experiences to the question of space commercialization. Specifically, I propose using these case studies to determine the applicability of analogies in the history of the United States to the present issues of building a space economy. Some examples may work better than others in specific settings, of course, but the probing questions about specific issues, problems, and solutions, and to specific issues, organizations, and individuals in this history, will help to illuminate possibilities for the future.

A Breathless Survey of American Spaceflight History

The dominant users of space from the beginning of the Space Age until the 1980s were governments, especially the two superpowers—the United States and the Soviet Union. There were several entities in the United States dedicated to space operations. On the civil side, NASA quickly emerged as the preeminent entity, but several parts of the national security apparatus also undertook space research, development, and operations. Later, such organizations as the National Oceanic and Atmospheric Administration (NOAA) entered the arena; by 2013, there were 14 separate U.S. government agencies with a role in space activities.[11]

NASA emerged in 1958 out of the "Cold War" rivalries of the United States and the Soviet Union. As the superpowers engaged in a broad contest over the ideologies and allegiances of the nonaligned nations of the world, space exploration was one major area contested. This served as the key that opened the door to aggressive space exploration, not as an end in itself, but as a means to achieving technological superiority in the eyes of the world. From the perspective of the 21st century, it is difficult to appreciate Americans' near-hysterical preoccupation with nuclear attack in the 1950s. Far from being the *Happy Days* of the television sitcom, the United States was a dysfunctional nation preoccupied with death by nuclear war. Schools required children to practice civil defense techniques and shield themselves from nuclear blasts—techniques, in some cases, as simple as crawling under their desks. Communities practiced civil defense drills, and families built personal bomb shelters in their backyards.[12] In the popular culture, nuclear attack was inexorably linked to the space above the United States, from which the attack would come.

The Soviets gained the upper hand in this competition on 4 October 1957, when they launched Sputnik, the first artificial satellite to orbit Earth, as part of a larger scientific effort associated with the International Geophysical Year (IGY).[13] While U.S. officials congratulated the Soviet Union for this

---

11 NASA, *Aeronautics and Space Report of the President, Fiscal Year 2008 Activities* (Washington, DC: NASA, 2009).

12 Elaine Tyler May, *Homeward Bound: American Families in the Cold War Era* (New York: Basic Books, 1988), pp. 93–94, 104–113.

13 There are several works on the history of Sputnik worthy of consideration. See Robert A. Divine, *The Sputnik Challenge: Eisenhower's Response to the Soviet Satellite* (New York: Oxford University Press, 1993); Walter A. McDougall, ...*The Heavens and the Earth: A Political History of the Space Age* (New York: Basic Books, 1985, rep. ed. Baltimore, MD: Johns Hopkins University Press, 1997); Rip Bulkeley, *The Sputniks Crisis and Early United States Space Policy: A Critique of the Historiography of Space* (Bloomington: Indiana University Press, 1991); Matt Bille and Erika Lishock, *The First Space Race: Launching the World's First Satellites* (College Station: Texas A&M University Press, 2004); Paul L. Dickson, *Sputnik: The Shock of the Century* (New York: Walker and Co., 2001); Matthew Brzezinski, *Red Moon Rising: Sputnik and the Hidden Rivalries That Ignited the Space Age* (New York: Times Books, 2007). Other aspects of this subject are recounted in Martin Collins, ed., *After Sputnik: 50 Years of the Space Age* (New York: Collins, 2007); Asif A. Siddiqi, *Sputnik and the Soviet Space Challenge* (Gainesville: University Press of Florida, 2003); James J. Harford, *Korolev: How One Man Masterminded the Soviet Drive to Beat America to the Moon* (New York: John Wiley & Sons, 1997); Philip Nash, *The Other Missiles of October: Eisenhower, Kennedy, and the Jupiters, 1957–1963* (Chapel Hill: University of North Carolina Press, 1997); Peter J. Roman, *Eisenhower and the Missile Gap* (Ithaca, NY: Cornell University Press, 1995); Kenneth Osgood, *Total Cold War: Eisenhower's Secret Propaganda Battle at Home and Abroad* (Lawrence: University Press of Kansas, 2006); Homer Hickam, *Rocket Boys* (New York: Delacorte, 1999); Constance M. Green and Milton Lomask, *Vanguard: A History* (Washington, DC: Smithsonian Institution Press, 1971).

accomplishment, clearly many Americans thought that the Soviet Union had staged a tremendous coup for the communist system at U.S. expense.

After an arms race with its nuclear component and a series of hot and cold crises in the Eisenhower era, coupled with the launching of Sputnik and Sputnik II in 1957, the threat of holocaust felt by most Americans and Soviets was now not just a possibility, but a seeming probability. For the first time, enemies could reach the United States with a radical new technology.[14] It was a shock, introducing the illusion of a technological gap and providing the impetus for the 1958 act creating NASA. Sputnik led directly to several critical efforts aimed at "catching up" to the Soviet Union's space achievements. Among them were these:

- A full-scale review of both the civil and military programs of the United States (scientific satellite efforts and ballistic missile development[).]
- Establishment of a Presidential Science Advisor in the White House who had responsibility for overseeing the activities of the Federal government in science and technology.
- Creation of the Advanced Research Projects Agency in the Department of Defense, and the consolidation of several space activities under centralized management.
- Establishment of NASA to manage civil space operations.
- Passage of the National Defense Education Act to provide federal funding for education in the scientific and technical disciplines.[15]

More immediately, the United States launched its first Earth satellite on 31 January 1958, when Explorer 1 documented the existence of radiation zones encircling Earth. Shaped by Earth's magnetic field, what came to be called the Van Allen Radiation Belt partially dictates the electrical charges in the atmosphere and the solar radiation that reaches Earth. It also began a series of scientific missions to the Moon and planets in the latter 1950s and early 1960s.[16]

Because of this perception, Congress passed and President Dwight D. Eisenhower signed the National Aeronautics and Space Act of 1958. This legislation established the National Aeronautics and Space Administration (NASA) with a broad mandate to explore and use space for "peaceful purposes for the benefit of all mankind."[17] The core of NASA came from the earlier National Advisory Committee for Aeronautics, with its 8,000 employees, an annual budget of $100 million, and its research laboratories.

---

14 See Roger D. Launius, John M. Logsdon, and Robert W. Smith, eds., *Reconsidering Sputnik: Forty Years Since the Soviet Satellite* (Amsterdam, The Netherlands: Harwood Academic Publishers, 2000).

15 Roger D. Launius, "Eisenhower, Sputnik, and the Creation of NASA: Technological Elites and the Public Policy Agenda," *Prologue: Quarterly of the National Archives and Records Administration* 28 (summer 1996): 127–143; Roger D. Launius, "Space Program," in Robert H. Ferrell and Joan Hoff, senior eds., *Dictionary of American History: Supplement* (New York: Charles Scribner's Sons Reference Books, 1996), 2:221–223.

16 See James A. Van Allen, *Origins of Magnetospheric Physics* (Washington, DC: Smithsonian Institution Press, 1983); Matthew J. Von Benke, *The Politics of Space: A History of U.S. Soviet/Russian Competition and Cooperation in Space* (Boulder, CO: Westview Press, 1997).

17 "National Aeronautics and Space Act of 1958," Public Law #85-568, 72 Stat., 426, signed by the President on 29 July 1958, Record Group 255, National Archives and Records Administration, Washington, DC; Alison Griffith, *The National Aeronautics and Space Act: A Study of the Development of Public Policy* (Washington, DC: Public Affairs Press, 1962), pp. 27–43.

It quickly incorporated other organizations into the new agency, notably the space science group of the Naval Research Laboratory in Maryland, the Jet Propulsion Laboratory managed by the California Institute of Technology for the Army, and the Army Ballistic Missile Agency in Huntsville, Alabama.[18]

The Soviet Union, while not creating a separate organization dedicated to space exploration, infused money into its various rocket design bureaus and scientific research institutions. The chief beneficiaries of Soviet spaceflight enthusiasm were the design bureau of Sergei P. Korolev (the chief designer of the first Soviet rockets used for the Sputnik program) and the Soviet Academy of Sciences, which devised experiments and built the instruments that were launched into orbit. With huge investments in spaceflight technology urged by Soviet premier Nikita Khrushchev, the Soviet Union accomplished one public relations coup after another against the United States during the late 1950s and early 1960s.[19]

In the United States, within a short time after NASA's formal organization, the new agency also took over the management of space exploration projects from other federal agencies and began to conduct space science missions—such as Project Ranger to send probes to the Moon, Project Echo to test the possibility of satellite communications, and Project Mercury to ascertain the possibilities of human spaceflight. Even so, these activities were constrained by a modest budget and a measured pace on the part of NASA leadership.

In an irony of the first magnitude, Eisenhower believed that the creation of NASA and the placing of so much power in its hands by the Kennedy administration during the Apollo program of the 1960s was a mistake. He remarked in a 1962 article: "Why the great hurry to get to the moon and the planets? We have already demonstrated that in everything except the power of our booster rockets we are leading the world in scientific space exploration. From here on, I think we should proceed in an orderly, scientific way, building one accomplishment on another."[20] He later cautioned that the Moon race "has diverted a disproportionate share of our brain-power and research facilities from equally significant problems, including education and automation."[21] He believed that Americans had overreacted to the perceived threat.

During the first 15 years of the Space Age, the United States emphasized a civilian space exploration program consisting of several major components:

- Human spaceflight initiatives—Project Mercury's single-astronaut program (flights during 1961–63) to ascertain if a human could survive in space, Project Gemini (flights during

18 Roger D. Launius, *NASA: A History of the U.S. Civil Space Program* (Malabar, FL: Krieger Publishing Co., 1994), pp. 29–41.

19 The standard works on this subject are Asif A. Siddiqi, *Challenge to Apollo: The Soviet Union and the Space Race, 1945–1974* (Washington, DC: NASA SP-2000-4408, 2000); James J. Harford, *Korolev: How One Man Masterminded the Soviet Drive To Beat America to the Moon* (New York: John Wiley & Sons, 1997).

20 Dwight D. Eisenhower, "Are We Headed in the Wrong Direction?" *Saturday Evening Post*, 11 August 1962, p. 24.

21 Dwight D. Eisenhower, "Why I Am a Republican," *Saturday Evening Post*, 11 April 1964, p. 19.

1965–66) to practice for space operations with two astronauts, and Project Apollo (flights during 1968–72) to explore the Moon.[22]

- Robotic missions to the Moon (Ranger, Surveyor, and Lunar Orbiter), Venus (Pioneer Venus), Mars (Mariner 4, Vikings 1 and 2), and the outer planets (Pioneers 10 and 11, Voyagers 1 and 2).[23]
- Orbiting space observatories (Orbiting Solar Observatory, Hubble Space Telescope) to view the galaxy from space without the clutter of Earth's atmosphere.[24]
- Remote sensing Earth satellites for information gathering (Landsat satellites for environmental monitoring).[25]
- Applications satellites such as communications (Echo 1, Television Infrared Observation Satellite [TIROS], and Telstar) and weather-monitoring instruments.[26]

---

22 NASA has documented well the history of the Mercury, Gemini, and Apollo programs since its first histories appeared in the 1960s. See James M. Grimwood, *Project Mercury: A Chronology* (Washington, DC: NASA SP-4001, 1963); Mae Mills Link, *Space Medicine in Project Mercury* (Washington, DC: NASA SP-4003, 1965); Loyd S. Swenson, Jr., James M. Grimwood, and Charles C. Alexander, *This New Ocean: A History of Project Mercury* (Washington, DC: NASA SP-4201, 1966; rep. ed. 1998); James M. Grimwood and Barton C. Hacker, with Peter J. Vorzimmer, *Project Gemini Technology and Operations: A Chronology* (Washington, DC: NASA SP-4002, 1969); Ivan D. Ertel and Mary Louise Morse, *The Apollo Spacecraft: A Chronology, Volume I, Through November 7, 1962* (Washington, DC: NASA SP-4009, 1969); Mary Louise Morse and Jean Kernahan Bays, *The Apollo Spacecraft: A Chronology, Volume II, November 8, 1962–September 30, 1964* (Washington, DC: NASA SP-4009, 1973); Courtney G. Brooks and Ivan D. Ertel, *The Apollo Spacecraft: A Chronology, Volume III, October 1, 1964–January 20, 1966* (Washington, DC: NASA SP-4009, 1973); Barton C. Hacker and James M. Grimwood, *On Shoulders of Titans: A History of Project Gemini* (Washington, DC: NASA SP-4203, 1977, rep. ed. 2002); Charles D. Benson and William Barnaby Faherty, *Moonport: A History of Apollo Launch Facilities and Operations* (Washington, DC: NASA SP-4204, 1978); Ivan D. Ertel and Roland W. Newkirk, with Courtney G. Brooks, *The Apollo Spacecraft: A Chronology, Volume IV, January 21, 1966–July 13, 1974* (Washington, DC: NASA SP-4009, 1978); Courtney G. Brooks, James M. Grimwood, and Loyd S. Swenson, Jr., *Chariots for Apollo: A History of Manned Lunar Spacecraft* (Washington, DC: NASA SP-4201, 1966; rep. ed. 1998); Roger E. Bilstein, *Stages to Saturn: A Technological History of the Apollo/Saturn Launch Vehicles* (Washington, DC: NASA SP-4206, 1980, rep. ed. 1997); Arnold S. Levine, *Managing NASA in the Apollo Era* (Washington, DC: NASA SP-4102, 1982); W. David Compton, *Where No Man Has Gone Before: A History of Apollo Lunar Exploration Missions* (Washington, DC: NASA SP-4214, 1989); Sylvia D. Fries, *NASA Engineers and the Age of Apollo* (Washington, DC: NASA SP-4104, 1992); Robert C. Seamans, Jr., *Aiming at Targets: The Autobiography of Robert C. Seamans, Jr.* (Washington, DC: NASA SP-4106, 1996); Glen E. Swanson, ed., *"Before This Decade Is Out…": Personal Reflections on the Apollo Program* (Washington, DC: NASA SP-4223, 1999); Richard W. Orloff, comp., *Apollo by the Numbers: A Statistical Reference* (Washington, DC: NASA SP-2000-4029, 2000); Robert C. Seamans, Jr., *Project Apollo: The Tough Decisions* (Washington, DC: NASA SP-2005-4537, Monographs in Aerospace History, No. 37).

23 On these missions, see the following NASA histories: R. Cargill Hall, *Lunar Impact: A History of Project Ranger* (NASA SP-4210, 1977); Homer E. Newell, *Beyond the Atmosphere: Early Years of Space Science* (NASA SP-4211, 1980); Edward Clinton Ezell and Linda Neuman Ezell, *On Mars: Exploration of the Red Planet, 1958-1978* (NASA SP-4212, 1984); John E. Naugle, *First Among Equals: The Selection of NASA Space Science Experiments* (NASA SP-4215, 1991).

24 Shami Chatterjee, "Exploring the Universe: The NASA Great Observatory Program," 2004 presentation in possession of author; NASA Education Division, *NASA Great Observatories* (Washington, DC: NASA EP-1998-12-384-HQ, 1998).

25 Pamela E. Mack, *Viewing the Earth: The Social Construction of Landsat* (Cambridge, MA: MIT Press, 1990).

26 Andrew J. Butrica, ed., *Beyond the Ionosphere: Fifty Years of Satellite Communication* (Washington, DC: NASA SP-4217, 1997).

- An orbital workshop for astronauts (Skylab).[27]
- A reusable spacecraft for traveling to and from Earth orbit (the Space Shuttle).[28]

The capstone of this effort was, of course, the human expedition to the Moon, Project Apollo. A unique confluence of political necessity, personal commitment and activism, scientific and technological ability, economic prosperity, and public mood made possible the 25 May 1961 announcement by President John F. Kennedy to carry out a lunar landing program before the end of the decade as a means of demonstrating the United States' technological virtuosity.[29]

Project Apollo, backed by sufficient funding, was the tangible result of an early national commitment in response to a perceived threat to the United States by the Soviet Union. NASA leaders recognized that while the size of the task was enormous, it was still technologically and financially within their grasp, but they had to move forward quickly. Accordingly, the space agency's annual budget increased from $500 million in 1960 to a high point of $5.2 billion in 1965. A comparable percentage of the $1.9 trillion federal budget in 2006 would have equaled more than $77 billion for NASA, whereas the Agency's actual budget then stood at $16.6 billion. NASA's budget began to decline beginning in 1966 and continued a downward trend until 1975. NASA's fiscal year (FY) 1971 budget took a battering, forcing the cancellation of Apollo missions 18 through 20. With the exception of a few years during the Apollo era, the NASA budget has hovered at slightly less than 1 percent of all money expended by the U.S. Treasury. Stability has been the norm as the annual NASA budgets have incrementally gone up or down in relation to that 1 percent benchmark (see figure 2).[30]

While there may be reason to accept that Apollo was transcendentally important at some sublime level, assuming a generally rosy public acceptance of it is at best a simplistic and ultimately unsatisfactory conclusion. Indeed, the public's support for space funding has remained remarkably stable at approximately 80 percent in favor of the status quo since 1965, with only one significant dip in support in the early 1970s. However, responses to funding questions on public opinion polls are extremely sensitive to question wording and must be used cautiously.[31] Polls in the 1960s consistently ranked spaceflight near the top of those programs to be cut in the federal budget. Most Americans seemingly preferred

---

27 Roland W. Newkirk and Ivan D. Ertel, with Courtney G. Brooks, *Skylab: A Chronology* (Washington, DC: NASA SP-4011, 1977); W. David Compton and Charles D. Benson, *Living and Working in Space: A History of Skylab* (Washington, DC: NASA SP-4208, 1983).

28 T. A. Heppenheimer, *The Space Shuttle Decision: NASA's Search for a Reusable Space Vehicle* (Washington, DC: NASA SP-4221, 1999); Clay Morgan, *Shuttle-Mir: The U.S. and Russia Share History's Highest Stage* (Washington, DC: NASA SP-2001-4225, 2001).

29 In addition to the above books on Apollo, see Edgar M. Cortright, ed., *Apollo Expeditions to the Moon* (Washington, DC: NASA SP-350, 1975); W. Henry Lambright, *Powering Apollo: James E. Webb of NASA* (Baltimore, MD: Johns Hopkins University Press, 1995); David West Reynolds, *Apollo: The Epic Journey to the Moon* (New York: Harcourt, 2002).

30 These observations are based on calculations using the budget data included in the annual *Aeronautics and Space Report of the President, 2003 Activities* (Washington, DC: NASA Report, 2004), appendix E, which contains this information for each year since 1959; NASA, "National Aeronautics and Space Administration President's FY 2007 Budget Request," FY 2007 Budget Request, 6 February 2006, p. I, NASA Historical Reference Collection.

31 Stephanie A. Roy, Elaine C. Gresham, and Carissa Bryce Christensen, "The Complex Fabric of Public Opinion on Space," IAF-99-P.3.05, presented at the International Astronautical Federation annual meeting, Amsterdam, The Netherlands, 5 October 1999.

Figure 2
NASA Budget as Percentage of Federal Budget

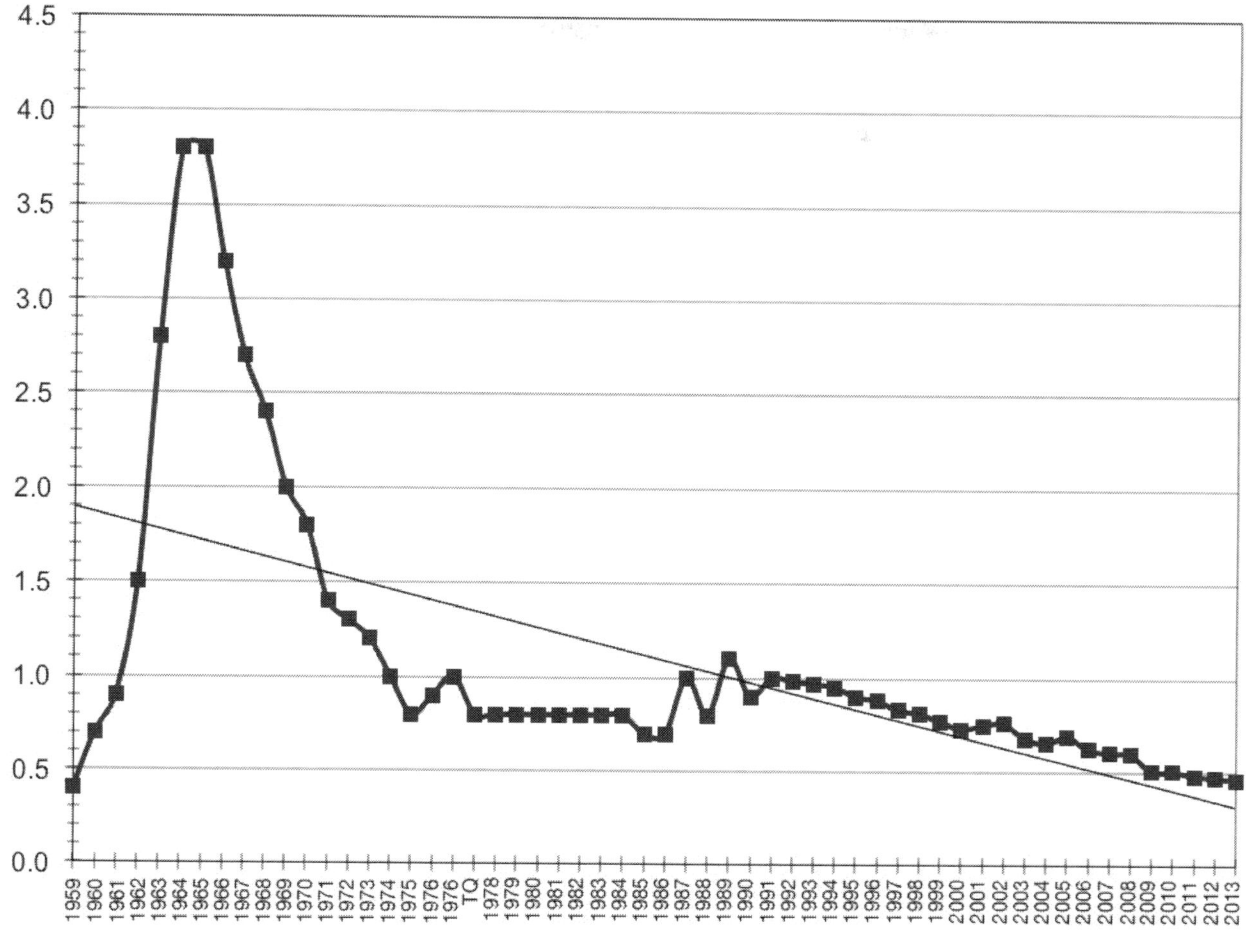

Source: Calculated from the *Aeronautics and Space Report of the President, Fiscal Year 2008 Activities* (Washington, DC: NASA, 2010), appendix D; NASA's FY 2009, 2010, 2011, 2012, 2013, and 2014 Budgets, available online at *http://www.nasa.gov/news/budget/*.

doing something about air and water pollution, job training for unskilled workers, national beautification, and poverty before spending federal funds on human spaceflight. The following year, *Newsweek* stated: "The U.S. space program is in decline. The Vietnam war and the desperate conditions of the nation's poor and its cities—which make space flight seem, in comparison, like an embarrassing national self-indulgence—have combined to drag down a program where the sky was no longer the limit."[32]

Nor did lunar exploration in and of itself create a groundswell of popular support from the general public. The American public during the 1960s largely showed hesitancy to "race" the Soviets to the Moon, as shown in figure 3. "Would you favor or oppose U.S. government spending to send astronauts to the moon?" these polls asked, and in virtually all cases a majority opposed doing so, even during the height of Apollo. At only one point, October 1965, did more than half of the public favor continuing human

32 *The Gallup Poll: Public Opinion, 1935–1971*, III: *1959–1971*, pp. 1952, 2183–2184, 2209; *New York Times*, 3 December 1967. *Newsweek* is quoted in *An Administrative History of NASA*, chap. II, p. 48, NASA Historical Reference Collection, NASA History Office, Washington, DC.

lunar exploration. In the post-Apollo era, the American public has continued to question the validity of undertaking human expeditions to the Moon.[33]

Figure 3

Should the Government Fund Human Trips to the Moon?

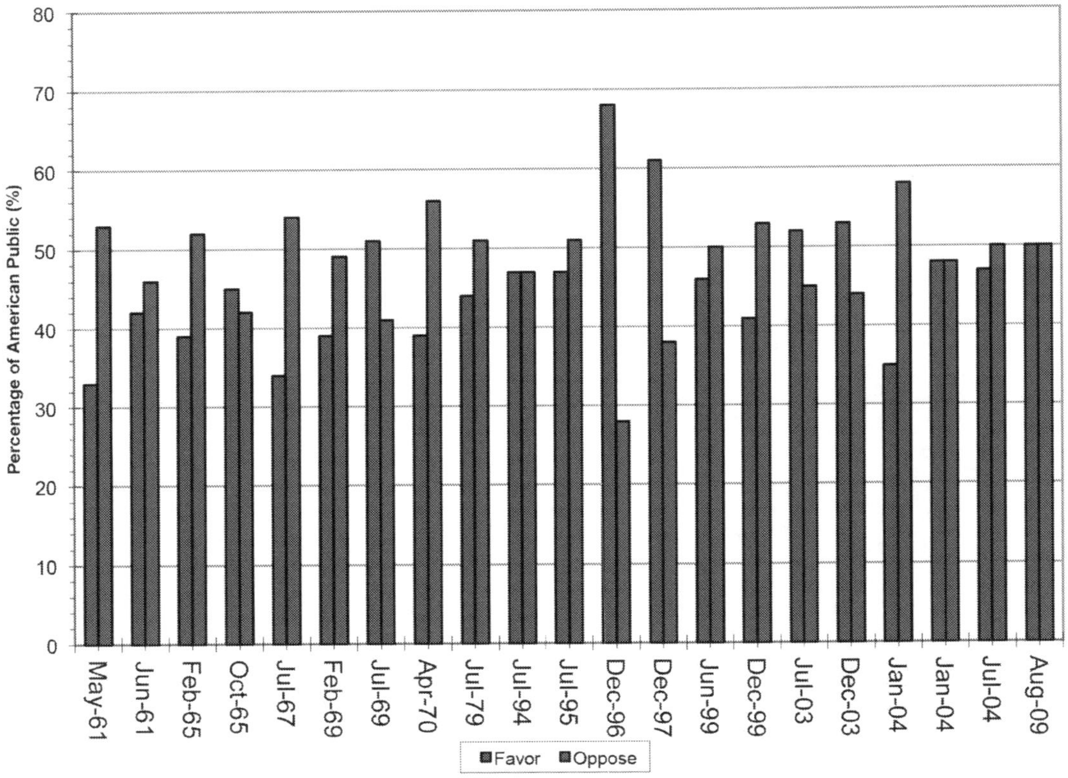

Source: Calculated from a set of Gallup, Harris, NBC/Associated Press, CBS/*New York Times*, and ABC/*USA Today* polls conducted throughout the 1960s, copies available in the NASA Historical Reference Collection, NASA History Program Office, Washington, DC.

Some conclude from these opinion polls that even though the American public might have been generally unsupportive of human lunar exploration, Project Apollo—wrapped as it was in the bosom of American virtue, advocated by the most publicly wholesome of astronaut heroes, and hawked by everyone from journalists to Madison Avenue marketers—enjoyed consistent popularity. There is some evidence to suggest this, but it is, on the main, untrue. From the 1960s to near the present, using the polling data that exist, there is little evidence to support public approval of an expansive lunar exploration and colonization program. One must conclude from hard evidence that the United States undertook and carried out Apollo not because the public clamored for it during the 1960s, but because it served other purposes. Furthermore, the polling data suggest that should the United States mount another human mission to the Moon in the future, it will also be because the mission serves a larger political, economic, or national defense agenda.[34]

33 This analysis is based on a set of Gallup, Harris, NBC/Associated Press, CBS/*New York Times*, and ABC/*USA Today* polls conducted throughout the 1960s, copies of which are available in the NASA Historical Reference Collection.

34 Roger D. Launius, "Public Opinion Polls and Perceptions of U.S. Human Spaceflight," *Space Policy* 19 (August 2003): 163–175.

These statistics do not demonstrate unqualified support for NASA's effort to reach the Moon in the 1960s. They suggest, instead, that the political crisis that brought public support to the initial lunar landing decision was fleeting, and within a short period the coalition that announced it had to retrench.[35] It also suggests that the public was never enthusiastic about human lunar exploration, especially about the costs associated with it. What enthusiasm it may have enjoyed waned over time until, by the end of the Apollo program in December 1972, one has the image of the program as something akin to a limping marathoner straining with every muscle to reach the finish line before collapsing.

The first Apollo mission of public significance was the flight of Apollo 8. On 21 December 1968, it took off atop a Saturn V booster from Kennedy Space Center. Three astronauts were aboard—Frank Borman, James A. Lovell, Jr., and William A. Anders—for a historic mission to orbit the Moon. After Apollo 8 made one and a half Earth orbits, its third stage began a burn to put the spacecraft on a lunar trajectory. It orbited the Moon on 24–25 December. The crew undertook a Christmas Eve broadcast, sending an image of Earth from lunar orbit while reading the first part of the Bible—"God created the heavens and the Earth, and the Earth was without form and void"—before sending Christmas greetings to humanity. The next day, they fired the boosters for a return flight, and they splashed down in the Pacific Ocean on 27 December. That flight was an enormously significant accomplishment because it came at a time when American society was in crisis over Vietnam, race relations, urban problems, and a host of other difficulties. And if only for a few moments, the nation united as one to focus on this epochal event. Two more Apollo missions occurred before the climax of the program, confirming that the time had come for a lunar landing.[36]

That landing came during the flight of Apollo 11, which lifted off on 16 July 1969 and, after confirmation that the hardware was working well, began the three-day trip to the Moon. Then, at 4:18 p.m. EST on 20 July 1969, the Lunar Module—with astronauts Neil Armstrong and Buzz Aldrin aboard—landed on the lunar surface while Michael Collins orbited overhead in the Apollo Command Module. After checkout, Armstrong set foot on the surface, telling millions who saw and heard him on Earth that it was "one small step for [a] man—one giant leap for mankind." Aldrin soon followed him out, and the two plodded around the landing site in the ⅙ lunar gravity; planted an American flag but omitted claiming the land for the United States, as had been routinely done during European exploration of the Americas; collected soil and rock samples; and set up scientific experiments. The next day, they launched back to the Apollo capsule orbiting overhead and began the return trip to Earth, splashing down in the Pacific on 24 July.[37]

Five more landing missions followed at approximately six-month intervals through December 1972, each of them increasing the time spent on the Moon. The scientific experiments placed on the Moon and

35 Roger D. Launius, "Kennedy's Space Policy Reconsidered: A Post–Cold War Perspective," *Air Power History* 50 (winter 2003): 16–29.

36 Robert Zimmerman, *Genesis: The Story of Apollo 8* (New York: Four Walls Eight Windows, 1998).

37 In addition to the many histories of Apollo that give center place to Apollo 11, there have been several memoirs by the astronauts on this mission. See Neil A. Armstrong, Michael Collins, and Edwin E. Aldrin, Jr., *First on the Moon: A Voyage with Neil Armstrong, Michael Collins and Edwin E. Aldrin, Jr.*, written with Gene Farmer and Dora Jane Hamblin (Boston: Little, Brown, 1970); Buzz Aldrin, *Return to Earth* (New York: Bantam Books, 1973); Michael Collins, *Carrying the Fire: An Astronaut's Journeys* (New York: Farrar, Straus and Giroux, 1974).

the lunar soil samples returned have provided grist for scientists' investigations ever since. The scientific return was significant, with the latter Apollo missions using a lunar rover to enhance substantially the ability to undertake scientific investigation. None of them, however, equaled the excitement of Apollo 11.

Project Apollo in general should be viewed as a watershed in the nation's history. It was an endeavor that demonstrated both the technological and economic virtuosity of the United States and established national preeminence over rival nations—the primary goal of the program when first envisioned by the Kennedy administration in 1961. At the same time, the Apollo program, while an enormous achievement, left a divided legacy for NASA and the aerospace community. The perceived "golden age" of Apollo created for the Agency an expectation that the direction of any major space goal from the president would always bring NASA a broad consensus of support and provide it with the resources and license to dispense them as it saw fit. Something most NASA officials did not understand at the time of the Moon landing in 1969, however, was that Apollo had not been conducted under normal political circumstances and that the exceptional circumstances surrounding Apollo would not be repeated.[38]

After Apollo—and the interlude of Skylab—the space program went into a holding pattern as nearly a decade passed. During that time it moved from its earlier heroic age to one that was more characterized as a "routinization" of activities, perspectives, and processes; it was an institutionalizing of critical elements from a remarkably fertile heroic time not at all unlike that analyzed by longshoreman philosopher Eric Hoffer in *The True Believer*.[39]

The Space Shuttle was intended to make spaceflight routine, safe, and relatively inexpensive. Although NASA considered a variety of configurations, some of them quite exotic, it settled on a stage-and-one-half partially reusable vehicle with an approved development price tag of $5.15 billion. On 5 January 1972, President Nixon announced the decision to build a space shuttle. He did so both for political reasons and for national prestige purposes. Politically, it would help a lagging aerospace industry in key states he wanted to carry in the next election, especially California, Texas, and Florida.[40] Supporters—*especially* Caspar W. Weinberger, who later became President Ronald Reagan's defense secretary—argued that building the shuttle would reaffirm America's superpower status and help restore confidence, at

38 As an example, see the argument made in George M. Low, Team Leader, to Richard Fairbanks, Director, Transition Resources and Development Group, "Report of the NASA Transition Team," 19 December 1980, NASA Historical Reference Collection, advocating strong presidential leadership to make everything right with the U.S. space program.

39 Eric Hoffer, *The True Believer: Thoughts on the Nature of Mass Movements* (New York: Harper & Row, 1951), pp. 3–23, 137–155. See also Max Weber, "The Pure Types of Legitimate Authority," in *Max Weber on Charisma and Institution Building: Selected Papers*, ed. S. N. Eisenstadt (Chicago: University of Chicago Press, 1968), p. 46.

40 George M. Low, NASA Deputy Administrator, Memorandum for the Record, "Meeting with the President on January 5, 1972," 12 January 1972, NASA Historical Reference Collection. The John Erlichman interview by John M. Logsdon, 6 May 1983, NASA Historical Reference Collection, emphasizes the political nature of the decision. This aspect of the issue was also brought home to Nixon by other factors, such as letters and personal meetings. See Frank Kizis to Richard M. Nixon, 12 March 1971; Noble M. Melencamp, White House, to Frank Kizis, 19 April 1971, both in Record Group 51, Series 69.1, Box 51-78-31, National Archives and Records Administration, Washington, DC.

home and abroad, in America's technological genius and will to succeed. This was purely an issue of national prestige.[41]

The prestige factor belies a critical component. United States leaders supported the Space Shuttle not on its merits, but on the image it projected. As a result, the Space Shuttle that emerged in the early 1970s was essentially a creature of compromise that consisted of three primary elements: a delta-winged orbiter spacecraft with a large crew compartment, a cargo bay 15 by 60 feet in size, and three main engines; two solid rocket boosters (SRBs); and an external fuel tank housing the liquid hydrogen and oxidizer burned in the main engines. The orbiter and the two solid rocket boosters were reusable. The Shuttle was designed to transport approximately 45,000 tons of cargo into low-Earth orbit, 115 to 250 statute miles above Earth. It could also accommodate a flight crew of up to 10 persons (although a crew of 7 would be more common) for a basic space mission of seven days. During a return to Earth, the orbiter was designed so that it had a cross-range maneuvering capability of 1,265 statute miles to meet requirements for liftoff and landing at the same location after only one orbit.[42]

After a decade of development, on 12 April 1981, Columbia took off for the first orbital test mission. It was successful, and after only the fourth flight in 1982, Reagan declared the system "operational." It would henceforth carry all U.S. government payloads; military, scientific, and even commercial satellites could all be deployed from its payload bay.[43]

The Shuttle soon proved disappointing. By January 1986, there had been only 24 Shuttle flights, although in the 1970s NASA had projected more flights than that for every year. Critical analyses agreed that the Shuttle had proven to be neither cheap nor reliable, both primary selling points, and that NASA should never have used those arguments in building a political consensus for the program. In some respects, therefore, many agreed that the effort had been both a triumph and a tragedy. The program had been engagingly ambitious and had developed an exceptionally sophisticated vehicle, one that no other nation on Earth could have built at the time. As such, it had been an enormously successful program. At the same time, the Shuttle was essentially a continuation of space spectaculars, à la Apollo, and its much-touted capabilities had not been realized. It made far fewer flights and conducted far fewer scientific experiments than NASA had publicly predicted.[44]

---

41 Caspar W. Weinberger, Memorandum for the President, via George Shultz, "Future of NASA," 12 August 1971, White House, Richard M. Nixon, President, 1968–1971 File, NASA Historical Reference Collection.

42 Alfred C. Draper, Melvin L. Buck, and William H. Goesch, "A Delta Shuttle Orbiter," *Astronautics & Aeronautics* 9 (January 1971): 26–35; Charles W. Mathews, "The Space Shuttle and Its Uses," *Aeronautical Journal* 76 (January 1972): 19–25; John M. Logsdon, "The Space Shuttle Program: A Policy Failure," *Science* 232 (30 May 1986): 1099–1105; Scott Pace, "Engineering Design and Political Choice: The Space Shuttle, 1969–1972," M.S. Thesis, MIT, May 1982; Harry A. Scott, "Space Shuttle: A Case Study in Design," *Astronautics & Aeronautics* 17 (June 1979): 54–58.

43 The standard work on the Shuttle and its operational history is Dennis R. Jenkins, *Space Shuttle: The History of the National Space Transportation System, the First 100 Missions* (Cape Canaveral, FL: Dennis R. Jenkins, 2001, 3rd edition).

44 Roger D. Launius, "The Space Shuttle—Twenty-five Years On: What Does It Mean To Have Reusable Access to Space?" *Quest: The History of Spaceflight Quarterly* 13, No. 2 (2006): 4-20.

This crescendo of criticism reached its loudest point following the loss of Challenger during launch on 28 January 1986. Although it was not the entire reason for the accident, the pressure to get the Shuttle schedule more in line with earlier projections throughout 1985 prompted NASA workers to accept operational procedures that fostered shortcuts and increased the opportunity for disaster. The accident, traumatic even under the best of situations, was made that much worse because Challenger's crewmembers represented a cross section of the American population in terms of race, gender, geography, background, and religion. The explosion became one of the most significant events of the 1980s, as billions around the world saw the accident on television and empathized with any one or more of the crewmembers killed.[45]

After the Challenger accident, the Shuttle program went into a two-year hiatus while NASA worked to redesign the system. The Space Shuttle finally returned to flight without incident on 29 September 1988.[46] Tragedy struck again when Columbia was lost during reentry on 1 February 2003. Successful Shuttle missions throughout the program's life included scientific and technological experiments, including the deployment of important space probes like the Magellan Venus radar mapper in 1989 and the Hubble Space Telescope in 1990, the flights of Spacelab, a dramatic three-person spacewalk in 1992 to retrieve a satellite and bring it back to Earth for repair, and missions visiting the Russian space station Mir and contributing to the orbital construction of an International Space Station.[47] With the end of the Space Shuttle Program in 2011, after 135 missions, the United States had no intrinsically national capability to reach orbit; it contracted for astronaut rides aboard Russian Soyuz capsules.

In 1984, as part of its interest in reinvigorating the space program, the Reagan administration called for the development of a permanently occupied space station. Although the station was first projected to cost $8 billion, within five years the projected costs had more than tripled and the station had become too expensive. NASA pared away at the station budget, and in the end the project was satisfactory to almost no one. In 1993, the international situation allowed NASA to negotiate a landmark decision to include Russia in the building of an International Space Station (ISS). By 1998, the first elements had been launched; in 2000, the first crew went aboard. At the beginning of the 21st century, the effort involving 16 nations was a shadow of what had been intended. It had been caught in the backwash of another loss of a Shuttle and the inability to complete construction and resupply the ISS. Consistently, the ISS has proven a difficult issue as policy-makers have wrestled with competing political agendas without consensus.[48]

As this situation went unresolved, on 14 January 2004, President George W. Bush announced a vision of space exploration that called for humans to reach for the Moon and Mars during the next 30 years. As stated

---

45 By far the best work on the Challenger accident is Diane Vaughan, *The Challenger Launch Decision: Risky Technology, Culture, and Deviance at NASA* (Chicago: University of Chicago Press, 1996).

46 The story of the return-to-flight effort is related in John M. Logsdon, "Return to Flight: Richard H. Truly and the Recovery from the *Challenger* Accident," in Pamela E. Mack, ed., *From Engineering Science to Big Science: The NACA and NASA Collier Trophy Research Project Winners* (Washington, DC: NASA SP-4219, 1998), pp. 345–364.

47 On the Columbia accident, see Roger D. Launius, "After Columbia: The Space Shuttle Program and the Crisis in Space Access," *Astropolitics* 2 (July–September 2004): 277–322.

48 A general history of space stations factual and fictional, planned and realized, is contained in Roger D. Launius, *Space Stations: Base Camps to the Stars* (Washington, DC: Smithsonian Books, 2003).

at the time, the fundamental goal of this vision was to advance U.S. scientific, security, and economic interests through a robust space exploration program. In support of this goal, the United States would

- implement a sustained and affordable human and robotic program to explore the solar system and beyond;
- extend human presence across the solar system, starting with a human return to the Moon by the year 2020, in preparation for human exploration of Mars and other destinations;
- develop the innovative technologies, knowledge, and infrastructures both to explore and to support decisions about the destinations for human exploration; and
- promote international and commercial participation in exploration to further U.S. scientific, security, and economic interests.

In so doing, the President called for the completion of the ISS and the retirement of the Space Shuttle fleet by 2010. Resources expended there would then be re-vectored toward creating the enabling technologies necessary to return to the Moon and eventually reach Mars. He also proposed a small increase in the NASA budget to help make this a reality. By the time of the 2008 presidential election, however, it had become a virtual certainty that the initiative would not be sustained in the next presidential administration. In the end, after years of lack of funding for Bush's "Vision for Space Exploration," there was little reason to believe that this vision would not follow the same path as the aborted Space Exploration Initiative (SEI), which had been announced with great fanfare in 1989 but derailed by the early 1990s.[49]

A combination of technological and scientific advancement, political competition with the Soviet Union, and changes in popular opinion about spaceflight converged in the 1950s to affect public policy in favor of government space exploration. This found tangible expression in efforts of the 1950s and 1960s to move forward with an expansive space program and the budgets necessary to support it. After that initial rise of effort, however, space exploration reached the equilibrium in the 1970s that it has sustained through the present. The American public is committed to a measured program that includes a modest level of human and robotic missions, Earth science activities, and technology development efforts. A longstanding fascination with discovery and investigation has nourished much of the interest by the peoples of the United States in spaceflight. By the end of the first decade of the 21st century, however, support for human space exploration has long been soft and there are no sustained efforts to return to the Moon or to go to Mars.

---

49 Frank Sietzen, Jr., and Keith L. Cowing, *New Moon Rising: The Making of the Bush Space Vision* (Burlington, Ontario: Apogee Books, 2004); Craig Cornelius, "Science in the National Vision for Space Exploration: Objectives and Constituencies of the 'Discovery-Driven' Paradigm," *Space Policy* 21 (February 2005): 41–48; Wendell Mendell, "The Vision for Human Spaceflight," *Space Policy* 21 (February 2005): 7–10.

Commercial Activities in Space

While much of the history of the Space Age is dominated by national actors, there has been, almost from the beginning, a significant private-sector involvement as well. The first commercial activities in space resulted from the initiatives of the telecommunications industry to extend operations beyond the atmosphere almost with the beginning of the Space Age. Indeed, satellite communication was the only truly commercial space technology to be developed in the first decade or so after Sputnik. Perhaps the first person to evaluate the technical and financial aspects of satellite communications was John R. Pierce of AT&T's Bell Telephone Laboratories, who in the mid-1950s outlined the utility of a communications "mirror" in space, estimating that such a satellite would be worth a billion dollars.[50]

The idea thrilled many space advocates, and with the dawning of the Space Age, the United States moved to exploit this opportunity. The first attempt was a NASA test program call Echo, which called for the orbiting of a 100-foot inflatable satellite covered with reflective material that scientists could bounce a radio beam off. Difficulties abounded in trying to launch an inflatable, passive satellite, but tests were finally successful on 12 August 1960.[51]

At the same time, Bell Telephone Laboratories pursued active-repeater communications satellites, the first of which was the Telstar project. In 1962, NASA contracted to launch the Bell-owned and -operated Telstar satellite, the first orbital device to relay television and telephone signals. In the years that followed, space-based telecommunications grew into a multibillion-dollar industry. Several generations of Telstars, as well as other types of communications satellites in Earth orbit, helped to make real-time global telecommunications a reality. By 1964, a range of satellites were operating—two AT&T Telstars, two Relays, and two Syncoms—and the global village promised by theorists was in the process of becoming. Moreover, the technological know-how for this first of all commercial activity in space had been transferred to companies other than AT&T. At the same time, live television broadcasts from the 1964 Tokyo Olympics provided to the public a glimpse of the dawning age of instantaneous global telecommunications.[52]

Seeing the enormous commercial potential of space-based communications, the U.S. Congress passed the Communications Satellite Act of 1962, creating a Communication Satellite Corporation (COMSAT), with ownership divided 50-50 between the general public and the telecommunications corporations to manage global satellite communications for the United States. Near the same time, U.S. leaders recognized the possibility of competition and participated in the establishment of the International Telecommunication Satellite Consortium (INTELSAT), with COMSAT as the U.S. manager, to pro-

50 James Schwoch, *Global TV: New Media and the Cold War, 1949–69* (Urbana: University of Illinois Press, 2009), pp. 140–141; "A Down-to-Earth Look at Space Communications," *Bell Telephone Magazine* 41/1 (spring 1962): 54; John R. Pierce, "Orbital Radio Relays," *Jet Propulsion*, April 1955, p. 44; J. R. Pierce, *The Beginnings of Satellite Communications* (San Francisco: San Francisco Press, 1968), pp. 9–12.

51 Donald C. Elder, *Out from Behind the Eight-Ball: A History of Project Echo* (San Diego, CA: Univelt, Inc., American Astronautical Society [AAS] History Series, Volume 16, 1995).

52 David J. Whalen, *The Origins of Satellite Communications, 1945–1965* (Washington, DC: Smithsonian Institution Press, 2002); Helen Gavaghan, *Something New Under the Sun: Satellites and the Beginning of the Space Age* (New York: Copernicus Books, 1998); Andrew J. Butrica, ed., *Beyond the Ionosphere: Fifty Years of Satellite Communication* (Washington, DC: NASA SP-4217, 1997).

vide an international communications satellite system. Founded by 19 nations, with eventual membership of well over a hundred, it was initially very much an American organization, with the United States controlling 61 percent of the voting authority and all the technology. This changed over time. INTELSAT oversaw the development of INTELSAT 1 in 1965, the first piece of the operational global communications satellite network. With this satellite system in orbit, the world became a far different place. Within a few years, telephone circuits increased from five hundred to thousands and live television coverage of events anywhere in the world became commonplace. On 6 April 1965, COMSAT's first satellite, Early Bird, was launched from Cape Canaveral.[53]

Although the initial launch vehicles and satellites were American, other countries had been involved from the beginning. By the time Early Bird was launched, the United Kingdom, France, Germany, Italy, Brazil, and Japan had established communications ground stations. From a few hundred telephone circuits and a handful of members in 1965, the INTELSAT system grew to embrace more members than the United Nations and to possess the technical capability to provide millions of telephone circuits. Global satellite communications had begun. Other companies expanded this system thereafter. Cost to carriers per circuit, and to individual customers, declined dramatically as the system matured. By the end of the century, orbiting satellites were generating billions of dollars annually in sales of products and services and had transformed global communication by facilitating commercial broadcasting, business and scientific exchanges, and telephone and Internet communication among individuals worldwide.[54]

But the core question to be considered in the context of space-based global telecommunications is whether or not this is more of a governmental activity than anything else. While AT&T developed the first communications satellite, the U.S. government launched it on a reconditioned military missile. While AT&T sought an open system for business, the government moved to regulate and control it. International space telecommunications followed a similar close relationship between government and industry. Accordingly, should satellite communications be viewed as a public trust or one that was a free-market arena? How should such activities, whatever the specific industry, be administered? These were large questions in modern American history, economics, and society. Additionally, the manner in which space enterprises were stimulated—investment, business models, returns on investment, and the like—has been a uniquely important topic for some time, but few have looked at how historical case studies might inform future efforts to stimulate space commerce. The story of the COMSAT Corp. is a case study of how government and the private sector undertook the development of what has become a remarkably lucrative space business. With the rise of a range of private-sector entrepreneurial firms interested in pursuing space commerce in the last two decades, the process whereby those might be incubated, fostered, and expanded comes to the fore as important public policy concerns as never before in the history of the Space Age.

---

53 Burton I. Edelson, "Global Satellite Communications," *Scientific American* (February 1977): 58–73; Burton I. Edelson and Louis Pollack, "Satellite Communications," *Science* (18 March 1977): 1125–1133; John McDonald, "The Comsat Compromise Starts as Revolution," *Fortune* (October 1965): 128–131, 202–212.

54 John Krige, Angelina Long Callahan, and Ashok Maharaj, *NASA in the World: Fifty Years of International Collaboration in Space* (New York: Palgrave Macmillan, 2013), chapter 5; John M. Logsdon, "The Development of International Space Cooperation," in John M. Logsdon, gen. ed., *Exploring the Unknown: Selected Documents in the Evolution of the U.S. Civil Space Program*, Volume II (Washington, DC: NASA SP-4407, 1996), pp. 7–10.

At sum, the story of satellite telecommunications is one of a mixed relationship. The government took the lion's share of the risk and managed carefully the development and fielding of the technology during the first 25 years of the Space Age. Only with changes in the structure of government involvement in the 1970s did the market loosen and the satellite communication industry achieve a form of takeoff, more than an order of magnitude of growth in less than a decade and significant expansion continuing beyond. Now, in the 21st century, COMSAT operators are to be found in almost every nation. The 1990s especially saw the rise of many Asian and African satellite operators, matching the growth seen in North America and western Europe in earlier decades. Actual hardware manufacture, however, has remained ensconced in the United States and Europe. Japan, India, China, and Russia have also built communications satellites, but their commercial viability is still undetermined.

From a few hundred telephone circuits and a handful of operators in the 1960s, this business has grown to a present-day system with millions of telephone circuits while costs to users have decreased immeasurably. Likewise, media telecommunications have been transformed through this capability. Western Union's Westar I, the first U.S. domestic communications satellite, was launched on 13 April 1974. As historian David Whalen has noted:

> In December of the following year RCA launched its own satellite, SATCOM F-1. In early 1976 AT&T and COMSAT launched the first of the COMSTAR series. These satellites were used for voice and data, but very quickly television became a major user. By the end of 1976 there were 120 transponders available over the U.S., each capable of providing 1,500 telephone channels or one TV channel. Very quickly the "movie channels" and "super stations" were available to most Americans. The dramatic growth in cable TV would not have been possible without an inexpensive method of distributing video.[55]

Into the 21st century, television still dominated domestic satellite communications, but data use has grown tremendously with the advent of new technologies making the transfer of data over such systems as the Internet possible.

A new age of space entrepreneurship in the United States really began with a set of decisions in the 1970s aimed at advancing commercial space activities. While economic growth and development, as well as international competitiveness, had long been goals of national space policy during the 1960s (with the emphasis on the space race), little had been accomplished to open space to commercial activities. This was reflected in a set of relatively narrow legislative and executive branch initiatives outlined in various laws and policy directives. Direct federal investment in space research and development (R&D) and technology served as the principal means of stimulating the commercial space community during that era.

For example, a 1977 NASA-funded study by the Hudson Institute made 100-year projections of space activities. This study was very unusual for NASA. It developed several scenarios detailing possible long-

55 Roger D. Launius, Erik M. Conway, Andrew K. Johnston, Zse Chien Wang, Matthew H. Hersch, Deganit Paikowsky, David J. Whalen, Eric Toldi, Kerrie Dougherty, Peter L. Hayes, Jennifer Levasseur, Ralph L. McNutt, Jr., and Brent Sherwood, "Spaceflight: The Development of Science, Surveillance, and Commerce in Space," *Proceedings of the IEEE* 100 (13 May 2012): 1785–1818.

term trends in space R&D, technology, defense, environmental, and economic development. Scenarios include both pessimistic and optimistic rates of growth and opportunities in space. Space was viewed as a resource and a place for economic enterprise. The study did not detail economic benefits but did make economic issues central to long-term space financing, exploration, and development. This is one of the first studies sponsored by NASA that gave equal weight to economic and technological trends and changes. The authors recognized that the purpose of the study was both to provide NASA with qualitative projections of the future for planning and public relations purposes and to provide very broad guidelines to answer questions about the value and long-term use of space.[56]

With the arrival of the Reagan administration in 1981, efforts to expand commercial activities in space became a much higher priority. These efforts yielded remarkable results. Accordingly, NASA undertook an in-depth legal and policy paper detailing the options available to NASA for stimulating commercial investments and opportunities. This study argued for a new way of thinking at NASA; as the authors noted, the Agency needed to consider commercial ventures rather than R&D for itself. It explicitly called for NASA to "more effectively encourage and facilitate private sector involvement and investment in civil space and space-related activities." To that end, "*NASA will redirect a portion of its space* research and development activities to assure that its R&D program supports the research, development and demonstration of space technologies with commercial application"[57] (emphasis in original). The study also recommended that NASA "explore new opportunities for the application of space technology and [improve] demonstrated technologies to achieve the extensive operational capabilities available today."[58] It sought to emphasize cost-sharing arrangements with NASA conducting R&D and the private sector undertaking operational activities. These efforts yielded remarkable results beginning in the 1990s (see figure 4).

At the behest of the White House, NASA created an Office of Technology Transfer to oversee this effort, which had some early successes, especially as NASA worked to alter the regulatory environment and incentivize investment. As Craig L. Fuller, a White House staffer for the Cabinet Council on Commerce and Trade, commented:

> The entrance of free enterprise into space for commercial activities conforms with national traditions. Private initiative has been the foundation of our nation's development and progress from its beginning. Even during the earliest explorations of the North American continent, explorers and pioneers were followed by traders and craftsmen who came to serve new settlements. Now, industrial entrepreneurs are following our astronauts into the new realms. Commercial expertise will perhaps do for space what the earliest American settlers did for our continent. They turned forbidding regions into prosperous and hospitable inhabited areas.[59]

---

56 William M. Brown and Herman Kahn, "Long-Term Prospects for Developments in Space (A Scenario Approach)," Hudson Institute, Inc., 30 October 1977, NASW-2924, pp. 257–274, NASA Historical Reference Collection.

57 "Space Commercialization Meeting," memo and agenda from the White House concerning a meeting on 3 August 1983.

58 Ibid.

59 Craig L. Fuller, "Memorandum for the Cabinet Council on Commerce and Trade," 10 April 1984, pp. 4–5, Space Policy Institute, George Washington University, Washington, DC.

Figure 4

Worldwide Commercial Space Revenues

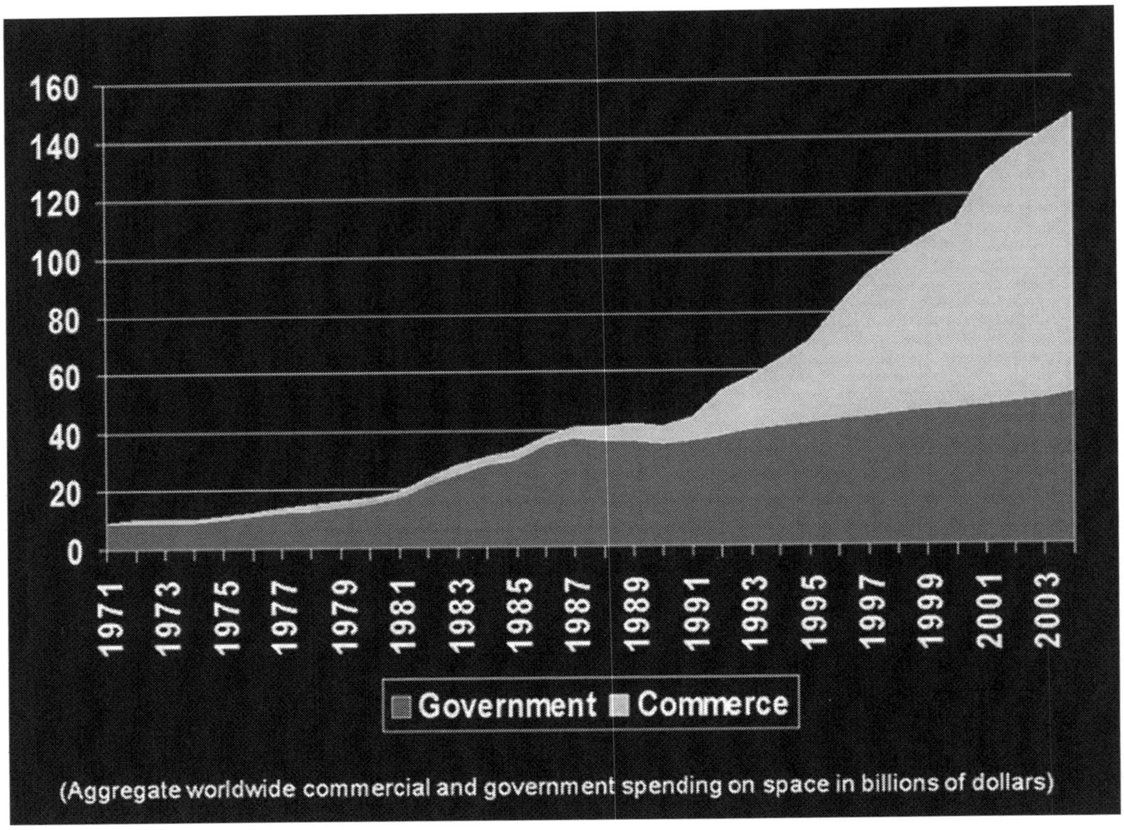

Source: Calculated from NASA, Aerospace Industry Association, and Facts on File data. Courtesy Roger D. Launius, Howard E. McCurdy, and Thomas Matula, "Historical Analogies for Space Commercialization," Executive Lunar Commerce Roundtable, Cox School of Business, Maguire Energy Institute, Southern Methodist University, Dallas, Texas, 24 June 2005.

Commercial space policy initiatives were aided by projections that space commerce could exceed $60 billion a year within 15 years. While highly controversial because of its optimism, this was still an influential study that sparked investment. Later revisions to the study offered more detail as well as more nuance to the analysis but still forecast a large rise in space commercial markets.[60]

In the last year of the Reagan administration, the Presidential Directive on National Space Policy offered, for the first time, a major section on commercial space efforts, reflecting positive commercial efforts in the communications satellite and launch vehicle sectors. The National Space Policy prohibited NASA from operating an expendable launch vehicle program and encouraged the government to purchase commercial launch services. It also called for open private space activities in microgravity, remote sensing, and other space ventures where there was potential for commerce.[61] The Bush administration issued its Commercial

60 *New York Times*, 24 June 1988, Section 3, p. 1; The Center for Space Policy, Inc., "Commercial Space Industry in the Year 2000, A Market Forecast," CSP Associates, Cambridge, MA, June 1985.

61 Office of the Press Secretary, White House, "The President's Space Policy and Commercial Space Initiative To Begin the Next Century," 11 February 1988, NASA Historical Reference Collection.

Space Policy Guidelines in 1991, which expanded on some of these same themes and emphasized the stimulation of private investment in launch, satellite, and support activities. The Guidelines explicitly recognized the use of space for commercial purposes and directed government agencies to procure every space technology available on the open market from private firms. It also mandated that NASA and other agencies using space avoid practices that might be viewed as deterring commercial space activities.[62]

The Clinton administration's 1996 Presidential Space Directive pushed these practices even further into the mainstream of national economic policy and international competitiveness. It explicitly stated: "To stimulate private sector investment, ownership, and operation of space assets, the U.S. Government will facilitate stable and predictable U.S. commercial sector access to appropriate U.S. Government space-related hardware, facilities and data." A continuation and extension of earlier policies going back to the Reagan administration, this directive was also the first to specifically detail the process by which the government might stimulate economic and business activity from space programs. It reflected the end of the Cold War, the shrinking federal discretionary budget, the maturity of some parts of the space program, and international competitive pressures.[63]

Collectively, this policy milieu stimulated some investment in such arenas as space launch and satellite development and operation. The communications satellite effort led the parade of investment—especially Iridium Satellite LLC, Globalstar, and DirecTV—but remote sensing, navigation, and other applications satellites followed. Such launchers as Conestoga and Orbital Sciences' Pegasus in the 1980s sought to alter the nature of space access. The only truly successful launch provider to come out of this period of ferment, however, was Orbital Sciences.[64]

Beginning in the mid-1990s, several startup companies were organized to develop new vehicles in response to the development of an expansive market for space access. Indeed, 1996 marked something of a milestone in the history of space access. In that year, worldwide commercial revenues in space for the first time surpassed all governmental spending on space, totaling some $77 billion. This growth continued in 1997, with 75 commercial payloads lofted into orbit and approximately 75 more military and scientific satellites launched. This represented a threefold increase over the number the year before. Market surveys for the period thereafter suggested that commercial launches would multiply for at least the next several years; one estimate held that 1,200 telecommunications satellites would be launched between 1998 and 2007.[65] In that context, many space launch advocates believed that the market had matured sufficiently that government investment in launch vehicle development was no longer necessary. Instead, they asked that the federal government simply "get out of the way" and allow the private sector to pursue development free from bureaucratic controls.[66]

---

62 National Space Policy Directive 3, "U.S. Commercial Space Policy Guidelines," White House, 12 February 1991, NASA Historical Reference Collection.

63 Office of the Press Secretary, White House, "National Space Policy, Fact Sheet," 19 September 1996, NASA Historical Reference Collection.

64 Roger D. Launius, "Between a Rocket and a Hard Place: The Challenge of Space Access," in W. Henry Lambright, ed., *Space Policy in the 21st Century* (Baltimore, MD: Johns Hopkins University Press, 2002), pp. 15–54.

65 Tim Beardsley, "The Way To Go in Space," *Scientific American* (March 1999, special issue on "The Future of Space Exploration").

66 Craig R. Reed, "Factors Affecting U.S. Commercial Space Launch Industry Competitiveness," *Business and Economic History* 27 (fall 1998): 222–236; Andrew J. Butrica, "Commercial Spaceports: Hitching Your Wagon to a VentureStar," *Space Times: Magazine of the American Astronautical Society* 37 (September/October 2000): 5–10.

This modern "gold rush" sparked several new corporations to muscle their ways into the tight conglomerate of launch vehicle companies. One of the farthest along and best financed of this new breed is Kistler Aerospace Corporation, based in Kirkland, Washington. Seeking low-cost access to space, Kistler employed Russian-built engines as a centerpiece of its K-1 reusable launcher. It was intended to deliver up to 10,000 pounds to orbit, depending on inclination. The first stage of this vehicle would fly back to the launch site; the second would orbit Earth before returning. Both stages would descend by parachute and land on inflatable air bags. Pioneer, Inc., has also been developing the Pathfinder rocketplane, an aerial propellant transfer spaceplane that accommodates a crew of two and can deliver a payload of 5,000 pounds to LEO. Kelly Space and Technology, Inc., has been at work on its Astroliner, a reusable spaceplane that can deliver 11,000 pounds to LEO for a cost of $2,000 per pound. Among these companies, the most interesting concept, albeit one that many people believe is doomed to failure, is the Roton by the Rotary Rocket Company. Roton was intended as a fully reusable, single-stage-to-orbit (SSTO) space vehicle designed to transport up to 7,000 pounds to and from LEO. Roton planned to enter commercial service in the year 2000 with a target price per flight of $7 million ($1,000 per pound).[67]

None of the myriad private efforts had yielded an operational space launcher by 2000. With the failure of the Iridium Corporation in spring 2000, a new satellite communications system that many believed would be the vanguard of a rapidly expanding space launch market, investment for space enterprises became even scarcer.[68] In some measure because of this, although they had previously eschewed government investment and the corresponding red tape, many of these startups began seeking capital from NASA and the military to support their efforts. Accordingly, it seemed that as the 21st century began, there was still a pressing need for substantial government investment in space launch R&D.

This became all the more significant with the 1 February 2003 loss of Space Shuttle Columbia and the recommendation to replace the Space Shuttle with another human-rated launcher. President George W. Bush announced his Vision for Space Exploration, challenging NASA to move beyond low-Earth orbit. What NASA came up with was the Constellation program, a presumed reuse of much of the existing Space Shuttle technology to build a new Ares I crew launch vehicle consisting of a Space Shuttle solid rocket booster as a first stage and an external tank as the beginning point for a second stage. A crewed space capsule, Orion, was to sit atop this system. A proposed second rocket, the Ares V cargo launch vehicle, would provide the heavy lift capability necessary to journey back to the Moon or to go beyond. Ares I was intended to carry a crew of up to six astronauts to low-Earth orbit in the Orion spacecraft, with the capability for expanding its use to send four astronauts to the Moon. Ares V was intended to serve as the Agency's primary vehicle for the delivery of large-scale hardware and cargo supporting an expansive space exploration agenda.[69]

---

67 Trevor C. Sorensen, AlliedSignal Technical Services Corp., "From the Earth to the Stars: The Future of Space Flight," presentation at the University of Kansas, 7 May 1999; Andrew J. Butrica, "The Commercial Launch Industry, Reusable Space Vehicles, and Technological Change," *Business and Economic History* 27 (fall 1998): 212–221.

68 Geoffrey V. Hughes, "The Iridium Effect," *Space Times: Magazine of the American Astronautical Society* 39 (March/April 2000): 23.

69 White House Press Release, "Fact Sheet: A Renewed Spirit of Discovery," 14 January 2004, available online at *http://www.whitehouse.gov/news/releases/2004/01/20040114-1.html*, accessed 4 April 2004.

Such was not the case when the Constellation Program was ended by President Barack Obama's administration because of cost and technical challenges in 2010. In a move that could be described by the great Monty Python line—"and now for something completely different"—President Obama's decision turned toward greater reliance on commercial space launch than ever before.[70] The new president convened a blue-ribbon panel—chaired by Norm Augustine, the former chief executive officer of Lockheed Martin and a longstanding space guru—which recommended in the fall of 2009 the harnessing of private-sector, especially entrepreneurial, firms in supporting Earth orbital operations instead of relying on Constellation. Augustine's report concluded: "Under current conditions, the gap in U.S. ability to launch astronauts into space will stretch to at least seven years. The Committee did not identify any credible approach employing new capabilities that could shorten the gap to less than six years." This would be true even with increased funding for NASA's program.[71]

The panel also noted that an increase of $3 billion per year for fiscal years 2010 to 2014 could return Constellation to health. Interestingly, there was no apparent enthusiasm in any corner of the political world for this type of expansion of the NASA budget. It was never considered as a serious solution to the Constellation Program's misfortunes on either side of the political spectrum. Some might suggest that this was because of the national debate on the deficit, which was ballooning exponentially at the time by more than one and a half trillion dollars, and everyone was looking for places to reduce the size of government. Such a hesitancy to increase funding for Constellation—something that had had to be done for the Space Shuttle Program by the Carter administration in the latter 1970s, when overruns had mired down the program and the promised 1978 first-flight date had been missed—suggested that something else might have been at work. Could it be that the national commitment to human spaceflight had waned to the extent that attrition seemed a likely way of ending it?

Regardless, the Augustine panel offered another option beyond that $15 billion increase for the existing Constellation Program, and this proved, ultimately, an attractive solution for political leaders of many stripes. "As we move from the complex, reusable Shuttle back to a simpler, smaller capsule, it is appropriate to consider turning this transport service over to the commercial sector," the panel concluded. "This approach is not without technical and programmatic risks, but it creates the possibility of lower operating costs for the system and potentially accelerates the availability of U.S. access to low-Earth orbit by about a year, to 2016. If this option is chosen, the Committee suggests establishing a new competition for this service, in which both large and small companies could participate."[72]

---

70 Major analyses of this transformation may be found in Marcia S. Smith, "President Obama's National Space Policy: A Change in Tone and a Focus on Space Sustainability," *Space Policy* 27 (February 2011): 20–23; John M. Logsdon, "Change and Continuity in U.S. Space Policy," *Space Policy* 27 (February 2011): 1–2; Gérard Brachet and Xavier Pasco, "The 2010 U.S. Space Policy: A View from Europe," *Space Policy* 27 (May 2011): 11–14; Yasuhito Fukushima, "An Asian Perspective on the New U.S. Space Policy: The Emphasis on International Cooperation and Its Relevance to Asia," *Space Policy* 27 (May 2011): 3–6; Sergey Avdeyev, Jean-Francois Clervoy, Jean-Marc Comtois, Takao Doi, Jeffrey Hoffman, Mamoru Mohri, and Gerhard Thiele, "Human Space Exploration—A Global Trans-Cultural Quest," *Space Policy* 27 (May 2011): 24–26.

71 Augustine Panel, "Seeking a Human Spaceflight Program Worthy of a Great Nation," 23 October 2009, available online at *http://www.nasa.gov/pdf/396093main_HSF_Cmte_FinalReport.pdf*, accessed 9 June 2011.

72 Ibid.

The response to this report from the space community was immediate. Some administration officials urged that the President cancel Constellation; others rallied to its defense. Edward Crawley, a Massachusetts Institute of Technology professor and a member of the Augustine panel, remarked that Ares I was suffering from technical issues that could only be overcome with more money and time. "It was a wise choice at the time," said Crawley, when asked about originating the program in 2005. "But times have changed…the budgetary environment is much tighter, and the understanding of the cost and schedule to develop the Ares I has matured."[73]

Based on these responses, President Obama proposed on 1 February 2010 (with more details added in a presidential speech on 15 April) a new path for future U.S. human spaceflight efforts. Central to this would be the termination of Constellation as a single entity; the continuation of certain technology developments, such as the Orion space capsule; renewed commitment to operations on the International Space Station until at least 2020; and the fostering of private-sector solutions to human spaceflight operations in LEO. Since this declaration, numerous high-profile spaceflight advocates have weighed in on both sides. In April 2010, Apollo astronauts Neil Armstrong (Apollo 11), Gene Cernan (Apollo 17), and Jim Lovell (Apollo 8 and Apollo 13) sent the President a letter warning that the proposed change to human spaceflight "destines our nation to become one of second- or even third-rate stature."[74] Proponents of the new strategy, among them Apollo 11 astronaut Buzz Aldrin, countered that the President's approach would return NASA to its roots as a research and development organization, while private firms operate space systems. Turning LEO over to commercial entities could then empower NASA to focus on deep space exploration, perhaps eventually sending humans to Mars or elsewhere.[75]

Although it might have been an ideal time to discuss the prestige trap and the overarching rationale for human spaceflight, the debate has largely been over maintaining a traditional approach to human spaceflight with NASA dominating the effort, owning the vehicles, and operating them through contractors. That was the method whereby America went to the Moon; it has proven successful over 50 years of human space exploration. Then there are those from the "new space" world that emphasize allowing private-sector firms to seize the initiative and pursue entrepreneurial approaches to human spaceflight. Advocates of the more traditional approach believe that the other side will sacrifice safety; advocates of the entrepreneurial approach criticize the forces of tradition by pointing out their large, over-budget, and under-achieving space efforts.

While these concerns are ever-present in the current U.S. debate over the future of human transportation into space, the place of commercial activities in this arena seems assured. There does not look to be any consensus in favor of undertaking a traditional approach to space access, one that NASA dominates both from a market standpoint and an ownership imperative. As space policy analyst Stewart Money commented:

---

73 Amy Klamper, "NASA in Limbo as Augustine Panel Issues Final Report," *Space News*, 23 October 2009, available online at *http://www.spacenews.com/article/nasa-limbo-augustine-panel-issues-final-report*, accessed 29 April 2014.

74 Michael Sheridan, "Neil Armstrong, James Lovell Call Obama's Plans for Space Exploration, NASA, 'Misguided'," *New York Daily News*, 14 April 2010.

75 Stephanie Condon, "Neil Armstrong vs. Buzz Aldrin Over Obama's Space Plans," CBS News, available online at *http://www.cbsnews.com/news/neil-armstrong-vs-buzz-aldrin-over-obamas-space-plans/*, accessed 29 April 2014.

> While the market for publically-funded commercial crew transport to ISS is clearly limited, it is becoming increasingly certain. NASA Administrator Charles Bolden's emphatic declaration in support of commercial crew may well signify a bellwether date in space history.... The most significant implication may be that other opportunities for leveraging the spacecraft initially offered for ISS cargo and crew duties can begin to emerge in earnest.... The fact that several of the proposed crew transport craft are promoted as being able to launch on more than one rocket indicates an encouraging shift in the way some aerospace companies are approaching the marketplace. It is something more akin to an aircraft manufacturer's role, an analogy Boeing was happy to make in the same announcement.[76]

As Money concluded: "Taken together, these developments, along with others, offer the tantalizing possibility of being the opening act in a new era of both public and private spaceflight to low-Earth orbit."[77]

Because of this directive from the Obama administration, NASA is pursuing efforts to replace the Space Shuttle through a multiphase space technology development program known as Commercial Crew Development (CCDev). Intended to stimulate development of privately operated crew vehicles that travel to LEO, its first phase offered a token sum of $50 million during 2010 to stimulate five American companies to undertake research and development of new human spaceflight concepts and technologies. In its second phase, contracts of $269 million were awarded to four firms in April 2011; the objective was to move toward the establishment of one or more orbital spaceflight capabilities on which NASA could purchase cargo and eventually crew space.[78] Additional phases are also being pursued, expanding the effort and the number of firms involved. The Commercial Crew integrated Capability (CCiCap, formerly CCDev 3) added to this effort in mid-2012 with a proposed set of Space Act Agreements because of Congressional funding reductions to the program for FY 2012.[79]

The awardees were all American firms, and their approaches ranged from lifting body to capsule spacecraft. Those companies—Blue Origin, Sierra Nevada Corporation, Space Exploration Technologies (SpaceX), and Boeing, with Paragon Space Development Corporation receiving some funding to develop an environmental control and life support system (ECLSS)—are envisioned as a vanguard that will be able to usher in a new era of innovative human spaceflight opportunities. As announced by Ed Mango, NASA's Commercial Crew Program manager: "The next American-flagged vehicle to carry our astronauts into space

---

76 Stewart Money, "Taking the Initiative: SLI and the Next Generation," *Space Review*, available online at *http://www.thespacereview.com/article/1784/1*, accessed 21 March 2011.

77 Ibid.

78 NASA Press Release 10-277, "NASA Seeks More Proposals on Commercial Crew Development," 25 October 2010, available online at *http://www.nasa.gov/home/hqnews/2010/oct/HQ_10-277_CCDev.html*, accessed 9 June 2011.

79 Frank Morring, Jr., "NASA Shifts CCDev Back to Space Act Procurement," *Aviation Week & Space Technology*, 15 December 2011, available online by subscription at *http://www.aviationweek.com/Article.aspx?id=/article-xml/awx_12_15_2011_p0-406939.xml*, accessed 16 May 2012; NASA Office of Procurement, "Announcement NASA-CCiCap, Commercial Crew Integrated Capability," 7 February 2012, available in the NASA Historical Reference Collection; NASA, "Commercial Crew Integrated Capability, Announcement No. NASA-CCiCap," February 2012, available in the NASA Historical Research Collection.

is going to be a U.S. commercial provider. The partnerships NASA is forming with industry will support the development of multiple American systems capable of providing future access to low-Earth orbit."[80]

This ferment of ideas and broad set of actions stimulated through the CCDev program suggest that the prestige trap of human spaceflight remains very much a part of the current human space exploration agenda at NASA. As the recent policy debate has illuminated, human spaceflight remains the priority despite compelling arguments against its continued centrality. The unsettled nature of the debate, which has now been going on for more than two years, may be interpreted as a signal that the prestige trap so long ago accepted as the raison d'être of NASA may be waning. Every serious observer of space policy has stated that he or she has never seen such a fight over next directions for NASA. While few credible commentators have challenged the endeavor altogether, a question to be considered is whether or not the prestige trap has finally been sprung and made it impossible to move forward with a shared vision for the future of humans in space. Instead, there is a cacophony of competing agendas being expressed, and an officially sanctioned way forward seeks to remove even the level of government support provided for the Space Shuttle Program from human space operations for the indefinite future. Further human spaceflight would rise or fall based on the ingenuity of its advocates to fashion activities that could become commercially self-supporting. One senior analyst recently commented that in 40 years of watching space policy, he has not seen such an unsettled situation.[81]

Without question, space commercialization has followed a path of struggle and diversion. Regardless, it has also followed a path forward—perhaps two steps forward, one step backwards, and three steps sideways with every development—but the change has been profound over the course of the history of the Space Age.

This brings to the fore the central element of this study, the manner in which historical analogs from the American past may inform current and future commercial activities in space.

80 Chris Bergin, "Four Companies Win Big Money via NASA's CCDEV-2 Awards," 18 April 2011, available online at *http://www.nasaspaceflight.com/2011/04/four-companies-win-nasas-ccdev-2-awards/*, accessed 9 June 2011.

81 John M. Logsdon, "A New U.S. Approach to Human Spaceflight?" *Space Policy* 27 (February 2011): 15–19.

The Use and Abuse of Historical Analogs

It is a truism that every person in every organization ever created uses history to make decisions on a daily basis. This is essentially accomplished through the use of analogy, suggesting that some issue under current scrutiny is like, may be compared to, or otherwise is related to some historical example. Understanding what happened in those past, analogous instances, therefore, serves a valuable purpose in considering what to do in the present. The difficulties of analogs, however, are that they are routinely poorly applied to considerations of policies, priorities, and decision-making that might effectively be informed by careful analog studies. Unfortunately, most uses of historical understanding are implicit, relying on personal anecdotes and employing faulty logic in the comparison. We have certainly seen this in the context of issues concerning the exploration of the space frontier since virtually the beginning of the Space Age. These range from analogs comparing modern cruise ship vacations and future space tourism to using the space race between the United States and the Soviet Union of the 1960s as an analog to predict a similar space race between the United States and China.

Central to the usefulness of this case study is the role of analogs in understanding current and future events. There is a long history of the use and abuse of analogs, offering perspectives on how they might be effectively employed in analysis of current challenges.[82] This study will employ analog analysis to a set of terrestrial colonization stories to reach broad conclusions that may inform future efforts. I intend to structure this set of case studies along similar lines to those developed in Richard E. Neustadt and Ernest R. May's classic text, *Thinking in Time: The Uses of History for Decision Makers*.[83] The methods employed were the fruit of several years' worth of classes taught by the authors in the Harvard Business School. They offered a structure that called for an analysis of each analog along three dimensions: What are the similarities with the present situation? What are the differences? And then, what are the implications of these similarities and differences? This framework can be productive in analyzing innovation and strategic surprise.[84]

Political scientist Francis Gavin has refined this approach, laying out five key concepts that promise more effective historical analysis and their application to current situations. These include understanding and investigating the applicability of 1) vertical history, 2) horizontal history, 3) chronological proportionality, 4) unintended consequences, and 5) policy insignificance.[85]

---

82 Analogs have been a recurrent topic in the contemporary philosophy of science especially, with the work of Mary B. Hesse in 1966 beginning the discussion. Her *Models and Analogies in Science* (Notre Dame, IN: University of Notre Dame Press, 1966) makes clear that analogies are integral to understanding scientific practice in general and scientific advancement in particular. Other important works on this subject include Paul F. A. Bartha, *By Parallel Reasoning: The Construction and Evaluation of Analogical Arguments* (New York: Oxford University Press, 2010); D. Gentner, K. J. Holyoak, and B. N. Kokinov, eds., *The Analogical Mind: Perspectives from Cognitive Science* (Cambridge, MA: MIT Press, 2001); K. D. Ashley, *Modeling Legal Argument: Reasoning with Cases and Hypotheticals* (Cambridge, MA: MIT Press, 1990); H. Kyburg and C. M. Teng, *Uncertain Inference* (Cambridge, U.K.: Cambridge University Press, 2001); M. Mitchell, *Analogy-Making as Perception* (Cambridge, MA: MIT Press, 1993).

83 Richard E. Neustadt and Ernest R. May, *Thinking in Time: The Uses of History for Decision Makers* (New York: Free Press, 1986).

84 Some efforts along these lines have been attempted in the context of spaceflight previously. See especially Bruce Mazlish, ed., *The Railroad and the Space Program: An Exploration in Historical Analogy* (Cambridge, MA: MIT Press, 1965).

85 Francis Gavin, "Five Ways To Use History Well," presentation at Long Now Foundation Lecture Series, San Francisco, CA, 12 July 2010, available online at *http://longnow.org/seminars/02010/jul/12/five-ways-use-history-well/*, accessed 6 April 2012.

- *Vertical history* focuses on understanding why events occurred in the past. This is a very standard task of historical investigation, and the best works published in the field all effectively present the whys of history and not just the hows.
- *Horizontal history* explores the linkage of events across space, whether geographical, cultural, economic, political, or otherwise.
- *Chronological proportionality* emphasizes the long-term consequences of events; as an example, understanding and applying which scraps of history concerning the Spanish experience in America that will be helpful in analog to the issue of space colonization. Instances universally hailed as significant may prove over time to be less important that initially thought.
- *Unintended consequences* present the challenge of applying an analog initially seen as useful but turning out to be a negative in the long run, or vice versa.
- *Policy insignificance* is the challenge of applying analogies without full appreciation that the analogs may be less useful than envisioned in the policy-making process.

These ideas, coupled with formal analog studies and historical perspectives from Neustadt and May, offer the methodological perspective to be applied to this current study.

Case Studies

The six historical case studies presented here investigate the manner in which the federal government—through its broad mandate "to provide for the general welfare" of the people of the nation—fostered, through a range of efforts, many different initiatives that contributed to the nation's economy, culture, and democratic status. Each of the six cases offers meaningful lessons in political culture and circumstances, legal precedent, methods of execution, and change over time. Many are considered positive stories of successful public involvement in the private sector; some offer cautionary tales of overreach and over-management; and some present lost opportunities that might have been more successful had circumstances been handled differently.

The following historical analogs inform this study:

1. The transcontinental railroad as a means of opening the American West—space transportation and the opening of cislunar space to commercial activities may be possible using related models to stimulate investment.

2. Support for the aerospace industry through legislation, appropriate regulation, and subsidies to grow a robust air transport capability—specifically exploring how best to harness these ideas to help foster a fully commercial space transportation industry that does not rely on government contracts for its profits.

3. The telephone industry's rise as a means of enhancing communication throughout the nation and worldwide, with the government overreaching in seeking to regulate this activity as a public utility—and how this might inform decisions about the development of a space communication network.

4. U.S. sponsorship of Antarctic stations to establish geopolitical presence and scientific efforts in Antarctica, gradually becoming less inherently governmental and more private over time—and how this might relate to lunar base support and a gradual transition from government activity to public-private efforts.

5. TVA, rural electrification, Henry Clay's American System, and other large public works projects—offering models for orbital, cislunar, and lunar infrastructure and economic development.

6. The National Park Service and preservation of scenic and cultural sites and how to balance the needs of preservation with development—offering lessons for fostering space tourism in sub-orbital, orbital, and lunar settings, an issue that has arisen specifically in the current Google Lunar XPRIZE and the need to ensure the integrity of landing sites on the Moon.

This study emphasizes the manner in which similar and related circumstances, priorities, economies, and other patterns have appeared in related instances. While much is known about the history of these earlier efforts in general, information about their details, especially their financing and political relations, have been largely unexplored in the context of space commercial activities. This study makes explicit these comparisons.

Developing the Transcontinental Railroad

In 1965, historian Bruce Mazlish edited the NASA-sponsored study *The Railroad and the Space Program: An Exploration in Historical Analogy*, seeking to understand the historical record of government stimulation of private-sector investment in infrastructure for the public good. The study team explored several specific episodes of American railroad history. It took this as its mission: "In all of these studies an effort will be made to move from the impact of the railroad in the specific area under consideration to an analogy with the possible space impact today in similar areas."[86] The original study offered an outstanding overview of early American railroad history and technology, but it failed to present a compelling analog comparable to spaceflight. One of the essayists in the volume, Thomas P. Hughes, offered this insight: "Wherever and whenever nature in her non-animal manifestations frustrates man in the pursuit of his objectives, there exists a technological frontier." He added that space exploration was a new case because it was a completely new arena of activity: "The most extreme result of technological frontier penetration is the creation of a man-made environment and the rendering of nature imperceptible."[87]

While the result was disappointing at the time, there remain lessons to be gained in exploring the historical analog of railroad building and operation in the 19th century and benefits from applying those lessons to an expansion of space exploitation. While many are familiar with the promotion of American transcontinental railroad construction through land grants, that was far from the sum total of public assistance. National, state, and local governments engaged in a range of other stimulative efforts to facilitate railroad development. These included tax breaks, investment credits, and otherwise favorable decisions supporting these business interests. Public support also involved, in some instances, direct subsidies for a time, monopolies not only on railroad operations but also in ancillary and even tertiary industries, as well as changes to regulations to ease requirements for labor, safety, and other factors. If there is any one conclusion in this survey, it is that there is a broad range of options that have been pursued in the past to stimulate investment in infrastructure—in this case, railroads—that have application for future space operations.

The first issue to be considered in investigating the development of railroads in the United States is the close relationship between the government and the transportation industry. This predated the invention

86 Earl P. Steveson, "Report of the CS," Records of the Academy (American Academy of Arts and Sciences), 1963/1964, pp. 150–151, quoted in Jonathan Coopersmith, "Great (Unfulfilled) Expectations: To Boldly Go Where No Social Scientist Has Gone Before," in Steven J. Dick, ed., *Remembering the Space Age: Proceedings of the 50th Anniversary Conference* (Washington, DC: NASA SP-2008-4703, 2008), p. 142.

87 Thomas P. Hughes, "A Technological Frontier: The Railway," in Bruce Mazlish, ed., *The Railroad and the Space Program: An Exploration in Historical Analogy* (Cambridge, MA: MIT Press, 1965), p. 53.

of the steam locomotive and the building of railroads, but it has been a persistent and impermeable aspect of the subject's history. Government involvement might be considered positive or negative—it might be viewed as an intrusion on the free-enterprise economy or as a help in securing the public welfare—depending on perspective. According to Jeffrey R. Orenstein:

> In the transportation field, especially, it started with the earliest road and canal building efforts, continued during the creation of the Interstate Commerce Commission in the nineteenth century, and intensified with the temporary nationalization of railroads during World War I, the government promotions for emerging transportation modes, and the rail quasi-nationalization of the present era. Railroads, particularly, have been major targets of both direct and indirect interventionist public transportation policies for over a century and a half.[88]

The result of this approach to rail transportation produced both public subsidies and regulation, sometimes alternating between the two without apparent rationality.[89] From the very beginning, the political economy of railroads entered into a codependent relationship with government. When held in creative tension, this could be a positive development, but it has not always maintained that balance.[90]

Prior to the 1830s, the U.S. government had waffled back and forth regarding direct construction of what were euphemistically titled "internal improvements," especially roads, harbors, and canals. The National Road from the mid-Atlantic seaboard to Illinois—now the route of U.S. highway 40—was a famous example of government investment in national infrastructure. Senator Henry Clay of Kentucky famously championed what he referred to as the "American System" to remake the nation into a modern state, and such investments in transportation systems were a direct result. When Andrew Jackson became President after the election of 1828, however, this philosophy met powerful opposition from a White House that believed in individualism and self-reliance. On 27 May 1830, Jackson vetoed the Maysville Road bill, which proposed a government subscription of $150,000 to a company building a 60-mile turnpike in Kentucky. Although the veto was partly the result of animosity toward Henry Clay, Jackson also described the bill as "a measure of purely local character...conferring partial instead of general advantages" and therefore not in the general interest of the nation.[91] But Jackson went further in his farewell address in 1837, announcing that his actions had "finally overthrown...this plan of unconstitutional expenditure for the purpose of corrupt influence."[92]

---

88 Jeffrey R. Orenstein, *United States Railroad Policy: Uncle Sam at the Throttle* (Chicago: Nelson-Hall, 1990), p. 1.

89 William Thoms, "Nationalization, No; Statelization, Yes," *Trains Magazine*, April 1985, pp. 44–48; Roy Sampson, Martin Farris, and David Shrock, *Domestic Transportation: Practice, Theory, and Policy* (Boston, MA: Houghton-Mifflin, 1985, 5th ed.).

90 Orenstein, *United States Railroad Policy*, p. 22.

91 Carter Goodrich, *Government Promotion of American Canals and Railroads* (New York: Columbia University Press, 1960), pp. 40–42.

92 "Farewell Address of Andrew Jackson," in Joseph L. Blau, ed., *Social Theories of Jacksonian Democracy* (New York: Hafner, 1947), p. 305.

That position remained national policy for several decades thereafter, although at the state and local level, considerable public money aided such construction projects. In essence, according to historian Carter Goodrich:

> The national government might aid transportation in various ways which returned no direct income, but it must refrain from building revenue-producing public works. It was not to construct roads and canals on which tolls were to be collected. It was not to subscribe to the stock of improvement companies. Most certainly, it was not to undertake the burden and responsibility of a scientifically planned system of national improvements.[93]

All else, for many Jacksonian Democrats, should be left either to private enterprise or to the will of the local and state governments. It was those local and state authorities that spurred the first investment in railroad construction in the 1830s and 1840s.

The first railways in the United States emerged from a succession of experiments undertaken by dreamers and visionaries in the first part of the 19th century. They emphasized private-sector investment and free-market capitalism, but also predatory operations.[94] Although it was a private enterprise, and the investors guarded their prerogatives on that score, there was always a significant public investment. As one scholar concluded: "Opposition of vested interests such as canals, and even farmers, and the difficulties of raising venture capital, on the one hand, coupled with the desire of many localities for improved transportation, on the other, led a number of states to embark upon railroad construction."[95]

The first railroad to use a steam engine, the South Carolina Canal and Rail Road Company, was chartered 19 December 1827 and began operations in downtown Charleston in February 1829 for hauling cotton bales. On 1 April 1830, a mile of double-tracked railroad entered into operation, and within three years, it had been extended to Hamburg, South Carolina, a distance of 136 miles from Charleston.[96] Investors in the railway saw an immediate market, bringing cotton from midstate plantations to Charleston for shipment to textile makers elsewhere. Even so, they were unable to compete in the free market with canals already providing this service.

Accordingly, they sought and received support from the state to ensure solvency. This took a variety of forms: 1) direct South Carolina appropriations to the company, making the state an investor in the railway; 2) waivers of a portion of the state's tariff on exports; 3) contracts to carry the state's mail; 4) state endorsement and backing of corporate bonds that ensured that they could be sold for face value; and 5) U.S. government indirect support for surveying and laying out the railroad because of

93 Goodrich, *Government Promotion of American Canals and Railroads*, p. 43.

94 Robert Sobel, *The Fallen Colossus* (New York: Weybright and Talley, 1977), chapters 1 and 5; J. Daughen and P. Benzen, *The Wreck of Penn Central* (Boston, MA: Little, Brown, 1971); Richard Saunders, *The Railroad Mergers and the Coming of Conrail* (Westport, CT: Greenwood Press, 1978).

95 Dudley F. Pegrum, T*ransportation: Economics and Public Policy* (Homewood, IL: Richard D. Irwin, Inc., 1963), p. 52.

96 Association of American Railroads, *American Railroads: Their Growth and Development* (Washington, DC: Association of American Railroads, 1956), pp. 5–6; Pegrum, Transportation, pp. 51–52; Samuel M. Derrick, *Centennial History of South Carolina Railroad* (Columbia, SC: State Publishing Company, 1933).

its use as a transportation means for military purposes.[97] Even so, debt and bankruptcy forced the original railroad company to reorganize and restructure in the 1840s. As one scholar concluded: "Some of the aid was in the form of a tariff reduction on iron, of banking privileges, of tax abatements, of grants of parcels of land by right of way, and of construction. State and local aid was more generally in stock subscription, donation of state bonds, loans, and endorsement of railroad bonds."[98]

The boom in railroad construction throughout the United States took place during the remainder of the antebellum period, also with considerable government stimulation of the industry. This took the form of state and local tax relief, project endorsement, public investment, and outright granting of subsidies. During the period before the Civil War, the states of Illinois, Pennsylvania, Indiana, Michigan, Virginia, and Georgia all built railroads as public works. In other states—notably Massachusetts, North Carolina, and Missouri—railroads became public commodities when the state took ownership of bankrupt lines. Most of these turned out poorly, and the states divested themselves of ownership by the time of the Civil War.[99]

Local governments also directly supported railroad construction, especially when the Panic of 1837 sucked considerable investment capital out of the financial system. A good case study of this direct support may be found in the rivalry of Troy and Albany, New York, who both wanted control of shipment between New York and the Midwest. The Erie Canal of 20 years earlier had been a boon to Albany, but Troy sponsored the building of the Schenectady and Troy Railroad at a cost of $700,000. After it was completed in 1842, the city operated it as a municipal activity and began to gain market share from the Erie Canal. Not to be outdone, the city fathers in Albany supported the construction of the Mohawk and Hudson Line with a $250,000 investment. Neither proved successful over the long run, and both were acquired by the New York Central Railroad in the 1850s. Overall, more than $1.2 million—some $5 billion in 2010 dollars—was invested in railroads by local and state governments in New York between 1837 and 1860.[100]

Increasingly, as the century progressed, the federal government also scratched the railroad building itch. This came largely because of the rising realization that a transcontinental railroad would become necessary in the coming decades. In most instances, government investment took the form of direct land grants to railroad companies. The first of these came when the U.S. government granted land to the state of Illinois in 1850 and it, in turn, granted it to the Illinois Central Railroad. This came as a result of the Land Grant Act of 1850, which provided 3.75 million acres of land to the states to support railroad projects. By 1857, 21 million acres of public lands had been transferred to railroads in the Mississippi River valley. Government land grants quickly followed along the same lines to the states of Mississippi and Alabama.[101] Regardless of these government investments, many of the efforts went bankrupt. "The belief that the mere presence

---

97 Ulrich Bonnell Phillips, *A History of Transportation in the Eastern Cotton Belt to 1860* (New York: Columbia University Press, 1908), pp. 132–220.

98 Kent T. Healy, *The Economics of Transportation in America: The Dynamic Forces in Development, Organization, Functioning and Regulation* (New York: Ronald Press Company, 1940), p. 105.

99 *Pegrum, Transportation, p. 54.*

100 Harry H. Pierce, *Railroads of New York: A Study of Government Aid, 1826–1875* (Cambridge, MA: Harvard University Press, 1953), pp. 116–118, 178–192; Goodrich, *Government Promotion of American Canals and Railroads*, p. 58–62.

101 K. Austin Kerr, "Railroad Policy," in Julian E. Zelizer, ed., *The American Congress: The Building of Democracy* (New York: Houghton Mifflin, 2004), p. 288.

of the Illinois Central would bring prosperity to the state, or that the railroad would actively bring it about," historian Robert L. Brandfon concluded, "was an illusion."[102]

This legislation pioneered a standard approach that the federal government followed thereafter: alternate even-numbered sections of 6 miles on either side of the proposed railways that the company could develop and sell. Since the land had not been attractive because of a lack of transportation—it had been available for sale to anyone for years at $1.25 per acre—the railroad now found a ready market for land useful to farmers. They were able to raise the price per acre to $2.50 and still find buyers. By the 1950s, about 8 percent of all railroad mileage in the United States had been constructed using money secured from federal land grants; some 131 million acres of public lands had been turned over to railroad companies for private use through this system.[103]

These efforts were nothing in comparison to the railroad construction that took place during the Gilded Age. Congress modified its approach for land grants with the Pacific Railroad Act of 1862 because of the very present need to build a transcontinental railroad. In this effort, lawmakers resolved to "do enough, and only enough, to induce capitalists to build the Pacific railway." Signed into law by President Abraham Lincoln on 1 July 1862, this act authorized extensive grants of public land along the rights of way and the issuance of 30-year government bonds at 6 percent. These subsidies went directly to the Union Pacific Railroad and Central Pacific Railroad to support the construction of a continuous transcontinental railroad from Council Bluffs, Iowa, to Sacramento, California.[104]

The act's most famous provisions dealt with the land grants. Section 2 provided each company with contiguous rights of way for their rail lines as well as all public lands within 200 feet on either side of the track. Section 3 granted an additional 10 square miles of public land for every mile of track laid except when running through cities or across rivers. It allocated this land as "five alternate sections per mile on each side of said railroad, on the line thereof, and within the limits of ten miles on each side." This turned out to be a most lucrative transfer of public property to the private sector, giving the favored companies a total of 6,400 square acres for each mile of track. By the time of the completion of the first transcontinental line in 1869, something approaching 175 million acres of public land had been transferred to the Union Pacific and the Central Pacific.[105] The central provisions included the following:

- Granted 20 sections of land for every mile of completed railway.
- Allowed railroads to use the value of the land as collateral for private loans.
- Provided subsidy bonds, essentially a second mortgage, to lend funds to railroad firms.
- Arranged for loans to be repaid largely by transportation revenues and land sales.

---

102 Robert L. Brandfon, "Political Impact: A Case Study of a Railroad Monopoly in Mississippi," in Mazlish, *The Railroad and the Space Program*, p. 186.

103 Association of American Railroads, *American Railroads*, pp. 7–13.

104 "An Act to Aid in the Construction of a Railroad and Telegraph Line from the Missouri River to the Pacific Ocean, and To Secure to the Government the Use of the Same for Postal, Military, and Other Purposes," 12 Stat. 489, 1 July 1862, available online at *http://www.cprr.org/Museum/Pacific_Railroad_Acts.html*, accessed 29 August 2013.

105 Ibid.

- Gave the government nonmonetary benefits (troop transport cost reductions).
- Increased returns by some 2 percent.

Throughout this process, the Union Pacific undertook construction westward from a point near Omaha, Nebraska; the Central Pacific headed eastward from Sacramento, California. The meeting point of the two lines on 10 May 1869 proved memorable. Collis P. Huntington of the Central Pacific and Grenville Dodge of the Union Pacific linked the tracks at Promontory Summit (also known as Promontory Point), Utah.[106]

Railroad expansion provided new avenues of immigration into the Great Plains and Rocky Mountain West. The railroads made money doubly by transporting people, goods, and commodities for a price and by selling portions of their land to arriving settlers at a handsome profit. Lands closest to the tracks, of course, drew the highest prices.

The national government also levied requirements on the transcontinentals. Those receiving federal assistance were required to transport U.S. mail, troops, and property at reduced rates. A mail discount rate of 20 percent and a 50 percent reduction in all other government transportation fees served national interests. This arrangement only ended with congressional action in 1940. Dudley F. Pegrum concluded about this development:

> Public aid to railroad development resulted in a very rapid expansion of the railroad network, which probably assisted in opening up the country more rapidly than would have been the case otherwise. At the same time it gave rise to abuses that were to have serious repercussions later on. The inevitable overbuilding and the extensive duplication of competitive lines created excess capacity that resulted in unsatisfactory earnings and financial failures that still plague the industry. Financial abuses flourished under public aid and the totally inadequate standards of financial responsibility of the period.[107]

Without the assistance of the U.S. government, railroad construction between 1860 and 1900 would certainly have proceeded at a less aggressive pace. At the end of the Civil War, only 45,000 miles of track had been laid. Between 1871 and 1900, another 170,000 miles were added to the nation's railroad system. Much, but not all, of this growth came as a result of the efforts to construct transcontinental railroads.[108] The investment of large startup costs—track surveying and construction, rolling stock acquisition, and support and logistics systems establishment, all before any revenue could accrue—meant

---

106 Bruce Clement Cooper, *Riding the Transcontinental Rails: Overland Travel on the Pacific Railroad, 1865–1881* (Philadelphia: Polyglot Press, 2005): pp. 1–15.

107 Pegrum, *Transportation*, p. 57.

108 George Rogers Taylor, *The Transportation Revolution, 1815-1860* (New York: Holt, Rinehart, and Winston, 1951); John F. Stover, *The Railroads of the South, 1865–1900: A Study in Finance and Control* (Chapel Hill: University of North Carolina Press, 1955); John F. Stover, *American Railroads* (Chicago: University of Chicago Press, 1961); Robert W. Fogel, *The Union Pacific Railroad: A Case in Premature Enterprise* (Baltimore, MD: Johns Hopkins University Press, 1960); Carter Goodrich, *Government Promotion of American Canals and Railroads, 1800–1890* (New York: Columbia University Press, 1960); Alfred Dupont Chandler, *Henry Varnum Poor: Business Editor, Analyst, and Reformer* (Cambridge, MA: Harvard University Press, 1956); Alfred Dupont Chandler, *The Railroads, The Nation's First Big Business: Sources and Readings* (New York: Arno Press, 1965).

that both private banks and entrepreneurs shied away from investment. Government investment, largely but not entirely through land grants, contributed to the success of four out of the five transcontinental railroads that were built in the period between the Civil War and 1900. "The total aid in monetary terms of private citizens and local, state, and federal governments was estimated by the Federal Coordinator of Transportation to have amounted to $1.4 billion," as reported by Dudley Pegrum.[109] In 2013 dollars, that would amount to more than $45 billion.

There is more relevance to this story for spaceflight than might be immediately apparent. The direct comparison of the public-private partnership that created the transcontinental railroads is the potential for government stimulation of the launch industry. The challenge is technological in the sense that new launchers are necessary for efficient operations, just as the laying of track and the acquisition and operation of rolling stock were critical to the transcontinental carriers of the 19th century. The similarities include the high startup costs associated with new, more-efficient launchers; the highly regulated operational environment; and the high risk/high return potential of the endeavor. The question before policy-makers, like the Congress of 1862, is how best to "do enough, and only enough, to induce capitalists to build" new space transportation systems.

In the railroad example, the forms of support included the following:

1. Land grants as a means of offering potential future revenue, tied to success in creating the railroad system.
2. Direct government appropriations to the company involved in the endeavor.
3. Waivers/modifications to taxes and other regulatory requirements.
4. Contracts for services once capability is demonstrated.
5. Government endorsement and backing of corporate bonds/assets.
6. Indirect support for related but supplemental elements of the railroad transportation system.[110]

In every case, these government initiatives were intended to leverage (and not replace) existing private funding, especially additional industry and venture capital.

There are those who believe that the federal government has been responsible for the stagnation present in 50 years of rocket technology. No question, after half a century, access to space remains a difficult challenge. The technical challenge of reaching space with chemical rockets—particularly the high costs associated with space launch, the long lead times necessary for scheduling flights, and the modest reliability of rockets—has demonstrated the slowest rate of improvement of all space technologies. All space professionals share a responsibility for addressing these critical technical problems. The overwhelming

109 Pegrum, *Transportation*, pp. 439–440.

110 Phillips, *A History of Transportation in the Eastern Cotton Belt to 1860*, pp. 132–220.

influence that space access has on all aspects of civil, commercial, and military space efforts indicates that it should enjoy a top priority for the 21st century.[111]

Of course, a key element in the spacefaring vision long held in the United States is the belief that inexpensive, reliable, safe, and easy spaceflight is attainable. Indeed, from virtually the beginning of the 20th century, those interested in the human exploration of space have viewed as central to that endeavor the development of vehicles for flight that travel easily to and from Earth orbit. The more technically minded recognized that once humans had achieved Earth orbit about 200 miles up, they had conquered the vast majority of the atmosphere and the gravity well and were then about halfway to anywhere else they might want to go.[112]

A central element in solving the current space access problem is to stimulate private-sector innovation—accomplished through public-private partnerships—to develop new, safe, reliable, and inexpensive rockets. The private sector cannot solve all problems as if by magic. At the same time, the U.S. government must relax its restrictions on the transfer of rocket technology to foster private-sector space launch innovation across national boundaries. That is an exceptionally tall order since space is overrun with dual-use technology that is critical to national security. The problem here, as John Krige has noted, is that "collaboration has worked most smoothly when the science or technology concerned is not of direct strategic (used here to mean commercial or military) importance." He added that as soon as a government feels that its national interests are directly involved in a field of R&D, it would prefer to protect these capabilities from proliferation. He also noted that the success of cooperative projects may take as their central characteristic that they have "no practical application in at least the short to medium term."[113]

In the end, the six points noted above of government involvement in 19th-century transcontinental railroads remain valid to some degree or another with the exception of land grants: there are none to offer in orbital space. However, there is the related right of companies developing effective space transportation systems to access government assets in space—especially the International Space Station. To those six points, we might add the following:

- Private financing supplemented with government loans.
- Property and patent rights granted to participating firms.
- Broadly construed revenues produced from transportation and other fees.

---

111 More than 1,000 space access studies have reached this conclusion over the last 40 years. See Roger D. Launius and Howard E. McCurdy, *Imagining Space: Achievements, Predictions, Possibilities, 1950–2050* (San Francisco: Chronicle Books, 2001), chapter 4; United States Congress, Office of Technology Assessment, *Launch Options for the Future: Special Report* (Washington, DC: Government Printing Office, 1984); Vice President's Space Policy Advisory Board, "The Future of U.S. Space Launch Capability," Task Group Report, November 1992, NASA Historical Reference Collection; NASA Office of Space Systems Development, *Access to Space Study: Summary Report* (Washington, DC: NASA Report, 1994).

112 This is the premise of G. Harry Stine, *Halfway to Anywhere: Achieving America's Destiny in Space* (New York: M. Evans and Co., 1996), a book that explores the historical path of launch vehicle development in the United States.

113 John Krige, "The Politics of European Collaboration in Space," *Space Times: Magazine of the American Astronautical Society* 36 (September–October 1997): 4–9.

Regardless, one must ask these critical questions in the context of developing new space transportation structures: "How important, in the final analysis, is cheaper access to space? Is it really the key to future growth of space activities?" This seems to be at the crux of what will go into any stimulation of private space transportation effort.

One of the central tenets of the new space community is that modern advances in technology and materials will allow inexpensive access to LEO. Unfortunately, this has not come to pass as yet. Current technological, economic, and regulatory realities combine to prohibit payload delivery to LEO for less than $1,000 per pound without significant changes to the current policy arena. Moreover, and this may be the core challenge for the future, no one has yet documented a clear, solid business model that would lead to a privately funded and operated space transportation system. Government customers are the major users of space transportation, not settlers on the American frontier homesteading land near the railroad. A market that could support the costs of creating such vehicles still seems far removed from the realities before the United States in the near term. Regardless, government investment has been significant to the present, and there may be some expansion in the future with continued adherence to the following points:

1. Government granting of use of publicly owned assets in low-Earth orbit, especially the International Space Station.
2. Direct government appropriations to the companies building space launchers, such as an expanded CCDev-type program.
3. Waivers for/modifications to taxes and other regulatory requirements.
4. Contracts for services once capability is demonstrated.
5. Government endorsement and backing of corporate bonds/assets.
6. Indirect support for related but supplemental elements such as range management, indemnification, and International Traffic in Arms Regulations (ITAR).
7. Technological knowledge transferred from government research organizations to private-sector firms developing revolutionary launch capabilities.

Beyond low-Earth orbit, especially in terms of a lunar transportation capability, might the government foster private-sector development through the creation of such a system? Another question to ponder: Is a privately developed lunar transportation corridor possible? The experience of the railroad suggests that it might be privately financed, although it would still need to be supplemented with government loans/bonds or other means of limiting private risk. Revenues produced from transportation fees could become a boon to the companies, just as they were for the transcontinental railroads of the 19th century. Property and patent rights could be granted to participating firms. There are, of course, challenges to this approach. Fundamental would be an overturning of the Outer Space Treaty of 1967 and the Moon Treaty—of the latter of which the U.S. is not a signatory—since there are no possibilities of lunar land grants in the current international treaty system. There might be opportunities short of outright ownership that would allow for the right to use land and extract minerals. Moreover, there might be a delta between the costs to be incurred and the value of future patents, thereby limiting large investments. Regardless, there are applicable ideas from the railroad experience that might be pursued.

Fostering the Aerospace Industry

During the first 40 years of the Air Age, government officials in the United States undertook a series of initiatives designed to foster the development of a commercial aerospace industry in private hands. Of course, even before the first flight of the Wright brothers on 17 December 1903, the United States government had been involved in the quest to fly with heavier-than-air vehicles. It contributed $100,000 toward the flying experiments of Smithsonian Institution Secretary Samuel P. Langley that failed in the fall of 1903.[114]

Since that first ill-fated investment, the United States government has recognized the importance of fostering aerospace development for issues of national security and economic viability. Over the years, this has taken place in three distinct and significant arenas, all of them contributing to the national air transportation objective. The first was the military aeronautics endeavor: employing aeronautical equipment and proficiency for the defense of the nation. By far this has been the largest outlay of federal spending on aerospace, funding basic research, development of new and ever-more-sophisticated weapons systems, and operational capabilities.

Second, the government has been fundamentally involved in fostering the research and development of air technologies principally through civilian agencies such as the National Advisory Committee for Aeronautics (NACA) and the National Aeronautics and Space Administration (NASA). Finally, the government has been intrinsically involved in the direction and regulation of commercial aerospace activities, both domestic and overseas, to facilitate air commerce and such aerospace operations as satellite communications. All of these actions represented a general industrial policy focused on developing a commercial air transportation industry.[115]

Such an approach recognizes that the overall health of the American aerospace industry is critical both for national security and for economic competitiveness. Even so, it is something of a truism to suggest that anything other than what has passed for aerospace policy in this nation has been both ad hoc and

---

114 Norriss S. Hetherington, "The Langley and Wright Aero Accidents: Two Responses to Early Aeronautical Innovation and Government Patronage," in Roger D. Launius, ed., *Innovation and the Development of Flight* (College Station: Texas A&M University Press, 1999), pp. 18–51.

115 Lewis M. Branscomb, "Toward a U.S. Technology Policy," *Issues in Science and Technology* 7 (summer 1991): 50–55. See also Lewis M. Branscomb, ed., *Empowering Technology: Implementing a U.S. Strategy* (Cambridge, MA: MIT Press, 1993); Lewis M. Branscomb and James H. Keller, eds., *Investing in Innovation: Creating a Research and Innovation Policy that Works* (Cambridge, MA: MIT Press, 1999); Bruce L. R. Smith and Claude E. Barfield, eds., *Technology, R&D, and the Economy* (Washington, DC: Brookings Institute, 1996); David C. Mowery and Nathan Rosenberg, *Paths of Innovation: Technological Change in 20th-Century America* (New York: Cambridge University Press, 1998); David C. Mowery and Nathan Rosenberg, *Technology and the Pursuit of Economic Growth* (New York: Cambridge University Press, 1989); Nathan Rosenberg, *Exploring the Black Box: Technology, Economics, and History* (Cambridge, MA: MIT Press, 1994).

expeditious.[116] Since the 1980s, it has become increasingly important to be more aggressive in ensuring this type of direct support as a matter of industrial policy because of incursions of foreign competitors into the aerospace sector and the direct subsidies they enjoyed from their governments. "In effect, the federal government was limping toward a sort of industrial policy," claimed Norman E. Bowie in 1994. "Since American industry was failing to invest in sufficient research and development to bring new products to market that could compete internationally, especially with the Japanese, the government provided public funds to universities to help move the fruits of basic research in to the marketplace."[117] This came as a direct result of foreign competition in the aerospace arena; as an example, one of the major factors energizing U.S. government subsidies was the necessity to respond to the ever-present and widespread subsidization of aerospace industrial development in Europe.

As early as 1908, Congress glimpsed something of the potential of aviation in the nation's defense, setting the stage for the most important and sustained United States government involvement in aerospace activities in the 20th century. Congress provided, in the Army Appropriations Act of 1908, for the procurement of one Wright "Flyer" by the Aviation Section of the Army Signal Corps. The military possibilities of this new environment, documented by Lieutenant Benjamin D. Foulois in a July 1908 staff paper to the Signal Corps School at Fort Leavenworth, suggested that in future conflicts, aircraft would be employed to limit "the strategic movement of hostile forces before they have actually gained combat." The possibility of aerial interdiction, coupled with the destruction of an enemy's means of producing war materiel through strategic bombing, ensured that the military would acquire some aircraft for combat missions. Foulois anticipated the striking changes in warfare that would spring from the use of a third dimension above Earth's surface for combat, transportation, and reconnaissance.[118] His ideas represented a merger of theoretical ideas with technological reality, and that merger represented a new way of war, the new high ground of the battlefield of the 20th century.

Even so, the pace of change was slow. As late as 1914, the United States stood 14th in total funds allocated by nations to military aviation, far behind even Bulgaria and Greece. To a very real extent, the result of Samuel P. Langley's lack of success—I hesitate to call it failure because aerospace R&D is, at a fundamental level, a process of trial and error, build-test-retest, that ultimately leads to an advancing of the state of the art—served to stunt the development of the airplane in America. Although the United

---

116 The problem of aerospace policy is related to the larger theme of industrial policy. For discussions of this issue, see Malcolm L. Goggin, ed., *Governing Science and Technology in a Democracy* (Knoxville: University of Tennessee Press, 1986); Manfred Stanley, *The Technological Conscience: Survival and Dignity in an Age of Expertise* (New York: Free Press, 1978); Sylvia Doughty Fries, "Expertise Against Politics: Technology as Ideology on Capitol Hill, 1966–1972," *Science, Technology, & Human Values* 8 (spring 1983): 6–15; David McKay, *Domestic Policy and Ideology: Presidents and the American State, 1964–1987* (New York: Cambridge University Press, 1989).

117 Norman E. Bowie, *University-Business Partnerships: An Assessment* (Lanham, MD: Rowman & Littlefield, Publishers, Inc., 1994), p. 19.

118 Benjamin D. Foulois and Carroll V. Glines, *From the Wright Brothers to the Astronauts: The Memoirs of B. D. Foulois* (New York: McGraw-Hill, 1968), p. 43. See also Eugene M. Emme, "The American Dimension," in Alfred F. Hurley and Robert C. Ehrhart, eds., *Air Power and Warfare: The Proceedings of the 8th Military History Symposium, United States Air Force Academy, 18–20 October 1978* (Washington, DC: Office of Air Force History, 1979), p. 57; Roger D. Launius, "A New Way of War: The Development of Military Aviation in the American West, 1908–1945," *Military History of the West* 25 (fall 1995): 167–190. On Foulois's career, see John F. Shiner, *Foulois and the U.S. Army Air Corps, 1931–1935* (Washington, DC: Office of Air Force History, 1983).

States invented the airplane, by the time of World War I it was obvious that the knowledge required to fly efficiently had moved offshore and resided in Europe. This was true for two reasons.

First, European governments, as well as industrial firms, tended to be more supportive of what might be called "applied research." As early as 1909, the internationally known British physicist Lord Rayleigh was appointed head of the Advisory Committee for Aeronautics; in Germany, Ludwig Prandtl and others had begun the sort of investigations that soon made the University of Göttingen a center of theoretical aerodynamics. Additional programs were soon under way in France and elsewhere on the continent. As Smithsonian Institution Secretary Charles D. Walcott wrote to Congress in 1915:

> As soon as Americans demonstrated the feasibility of flight by heavier-than-air machines, France took the matter up promptly, and utilized all the available agencies, including the army, navy, and similar establishments, both public and private. Large sums were devoted to the research work by wealthy individuals, and rapid advance was made in the art. Germany quickly followed, and a fund of one million seven hundred thousand dollars was raised by subscription, and experimentation directed by a group of technically trained and experienced men.[119]

Walcott added that England, Germany, and Russia followed suit, leading the way into the air age. He noted that when World War I began in 1914, about 1,400 military aircraft existed, of which only 23 belonged to the United States.

Second, fueled by military necessity, the nations of Europe invested heavily in aeronautical technology and built flying machines of great complexity and significant capability, capability far outstripping anything that the United States could accomplish in the mid-1910s.[120] As a result, the small, fast, maneuverable, and heavily armed fighter aircraft emerged as a major component of the World War I battlefield. Although powered flight had been possible since 1903, as late as 1914 there was little understanding of what might be possible in warfare by extending it into three dimensions with the use of the airplane. European combatants on both sides transformed airplanes into "warplanes," evolving these vehicles through five essential generations during the Great War. Each stage represented a major technological breakthrough and was dominated by one side of the belligerents. It also forced the development of fighter tactics to make aerial combat more effective. In turn, each stage was made obsolete by its successor, and while vestiges of aircraft types and tactics might remain throughout the rest of the war, they became less significant as later developments passed them by.[121]

---

119 Charles D. Walcott, Secretary of the Smithsonian Institution, to Senator Benjamin R. Tillman, chairman of the Committee on Naval Affairs, "Memorandum on a National Advisory Committee for Aeronautics," 1 February 1915, reprinted in Alex Roland, *Model Research: The National Advisory Committee for Aeronautics, 1915–1958* (Washington, DC: NASA SP-4103, 1985), 2:594–595.

120 John H. Morrow, Jr., *German Air Power in World War I* (Lincoln: University of Nebraska Press, 1982), pp. 3–13; John H. Morrow, Jr., *The Great War in the Air: Military Aviation from 1909 to 1921* (Washington, DC: Smithsonian Institution Press, 1993).

121 Richard P. Hallion analyzes this transformation in his masterful *Rise of the Fighter Aircraft, 1914–1918* (Baltimore, MD: Nautical and Aviation Press, 1984).

Similar progress in the United States was slow in coming. Aware of European activity, Secretary Walcott of the Smithsonian obtained funds to dispatch two Americans on a fact-finding tour overseas. Albert F. Zahm taught physics and experimented in aeronautics at Catholic University in Washington, DC, while Jerome C. Hunsaker, a graduate of the Massachusetts Institute of Technology, was developing a curriculum in aeronautical engineering at the institute. Their report, submitted to Congress early in 1915, emphasized the galling disparity between European progress and American inertia. The visit also established European contacts that later proved valuable to the NACA.[122]

Congress began a buildup of aeronautical capability and created a permanent Aviation Section of the War Department as Europe descended into a treacherous conflict. When the U.S. entered World War I in April 1917, this process accelerated and the government made significant investments in the aviation industry and expanded the procurement of military aircraft from 350 on order to an ambitious program to develop and produce 22,000 modern military aircraft by July 1918. Even without achieving this goal—U.S. manufacturers delivered 11,950 planes to the government during the war—the massive military appropriations gave the nascent aviation industry a huge boost. Equally important, the infrastructure of military aviation was solidified, and by 1919, the Army Air Service had established 69 airfields in the United States. All of these bases became part of a nationwide network of airways and landing fields that permitted rapid movement of units across the country for military purposes.[123]

Although there was a lull in aeronautical interest for the military following World War I and expansion slowed to a trickle, the military aspects of aviation would not go away. The amount of funding for military aviation declined every year by more than 10 percent after 1918 until it reached a low of $12.6 million in 1924. Many people questioned these government expenditures virtually as soon as the war was won. In response, Glenn L. Martin commented:

> Only a failure of the United States government to place orders with our successful airplane designers and builders will cause our aircraft industrial strength to slip back into the position it occupied three years ago. A vital point is being overlooked by the American people. It is immediately evident that the industrial strength of the United States must be at the war strength all the time.... The government must stimulate and aid in the application of aircraft industrially, and also aid in foreign trade, furnishing sufficient outlet for industrial aviation and guaranteeing a continuity of production at the required rate.[124]

Martin was right when he said this in 1920. He complained that the government required a strong aerospace industry as a guarantee of national defense and should put money into it as a matter of indus-

122 Walcott to Tillman, "Memorandum on a National Advisory Committee for Aeronautics," 1 February 1915, reprinted in Roland, *Model Research*, 2:593–597. See also William F. Trimble, *Jerome C. Hunsaker and the Rise of American Aeronautics* (Washington, DC: Smithsonian Institution Press, 2002).

123 On the U.S. effort in the First World War, see Charles J. Gross, *American Military Aviation: The Indispensable Arm* (College Station: Texas A&M University Press, 2002), pp. 26–47.

124 Quoted in Lt. Col. Ellen M. Pawlikowski, "Surviving the Peace: Lessons Learned from the Aircraft Industry in the 1920s and 1930s," Thesis, Industrial College of the Armed Forces, National Defense University, Fort McNair, Washington, DC, 1994, p. 1.

trial policy. A major breakthrough came with the Army Air Corps Act of 1926, which renamed the Air Service as the Army Air Corps, provided for an Assistant Secretary of War for Air and mandated a five-year Air Corps expansion program.[125]

As the United States began to rearm in the latter 1930s, the nation's leaders first officially recognized that the strength of the Army's air arm was critical and found it woefully inadequate. General George C. Marshall recalled that it "consisted of a few partially equipped squadrons serving the continental United States, Panama, Hawaii, and the Philippines; their planes were obsolescent and could hardly have survived a single day of modern aerial combat."[126] Harry Hopkins, President Franklin D. Roosevelt's confidant, commented shortly after Pearl Harbor that "[t]he President was sure that we were going to get into the war and believed that air power would win it."[127] Because of these inadequacies, in 1934 Congress appropriated $23.3 million for the use of the Army Air Corps, 8.4 percent of all Army appropriations. In 1936, Congress funded the construction of another wind tunnel at Langley and the lengthening of a tank used for seaplane research. It provided the impetus for additional funding through a special "Deficiency Appropriation Act" to fund the construction of new facilities, all because of war sentiment in Europe. In 1938, Roosevelt suggested that the Air Corps was operating with what could be politely called "antiquated weapons" and advocated increasing its strength to 30,000 airplanes from only a few hundred outdated biplanes.[128] In April 1939, when Congress passed the National Defense Act of 1940, it authorized the Army Air Corps to develop and procure 6,000 new airplanes, to increase personnel to 3,203 officers and 45,000 enlisted, and to spend $300 million. As a result, the Army Air Forces received $70.6 million, 15.7 percent of the Army's direct appropriations. This was only the beginning of a massive wartime expansion of the United States Army Air Forces during World War II.[129]

Virtually every study speaks of devastation to the aerospace sector of the U.S. economy that came with demobilization following World War II. From late 1943 on, the Joint Chiefs of Staff were sure of eventual victory and began to trim defense contracts for aircraft and war materiel. The result was that 1944 became the peak year of production, with 95,272 aircraft delivered, with another 48,912 delivered in the last year of the war. This was nearly half of the total of 316,495 aircraft produced during World War II. The next year, production slipped to just over 36,000 aircraft, but the vast majority of these were

125 On this subject, see James P. Tate, *The Army and Its Air Corps: Army Policy Toward Aviation, 1919–1941* (Maxwell Air Force Base [AFB], AL: Air University Press, 1998); Robert P. White, *Mason Patrick and the Fight for Air Service Independence* (Washington, DC: Smithsonian Institute Press, 2001); Harry Howe Ransom, "The Air Corps Act of 1926: A Study of the Legislative Process," Ph.D. Diss., Princeton University, 1953.

126 U.S. Army Chief of Staff, *Biennial Report, July 1, 1943–June 30, 1945* (Washington, DC: Government Printing Office, 1945), p. 117.

127 Quoted in Robert E. Sherwood, *Roosevelt and Hopkins: An Intimate History* (New York: Harper and Brothers, 1950 ed.), p. 100.

128 Report of the Secretary of War, FY 1938, pp. 26–27, U.S. Air Force (USAF) Historical Research Center, Air University, Maxwell AFB, AL; *Congressional Record*, 76th Cong., 1st Sess., p. 219.

129 Wesley Frank Craven and James L. Cate, eds., *The United States Air Force in World War II*, 6 vols. (Chicago, IL: University of Chicago Press, 1948), 1:104, 6:171–173; "Some Important Facts Regarding Expansion of NACA Research Facilities and War-time Status of NACA," 17 January 1946, NASA Historical Reference Collection; A. Hunter Dupree, *Science in the Federal Government: A History of Policies and Activities to 1940* (Cambridge, MA: Harvard University Press, 1957), p. 363.

commercial aircraft purchased after years of waiting for newer models. In 1947, production declined by more than half of its 1946 level.[130]

The postwar period brought sweeping organizational changes. The National Security Act of 1947 abolished two cabinet-level organizations, the Departments of War and the Navy, and established in their place the Department of Defense, transforming the air arm into the United States Air Force, coequal with the Army and the Navy. Since that time, several acts have followed that reformed various aspects of the Department of Defense and the aerospace infrastructure but left the basic institutional arrangements intact.[131]

Without question, the Cold War precipitated a continuation of the expansion of military aerospace activities. The military air and space component in the Cold War also involved a broad range of activities. The development, training, equipping, and employment of aerospace military power have extended from aircraft to missiles to satellites to other systems of both a passive and an active nature. Much of this has been carried out in a highly classified environment, such as satellite reconnaissance, with neither details nor records of government available for ready inspection. All have been justified as a means of maintaining the integrity of the nation against an aggressive communist menace.[132]

Post–Cold War national security concerns since 1990 have brought a new challenge to the American military aerospace sector. Since the collapse of the Soviet Union, a different set of priorities has replaced the powerful secular ideologies of democracy, communism, nationalism, fascism, and socialism that dominated international politics since the Enlightenment. These were not so much new priorities as ancient traditions based on ethnic, religious, kinship, or tribal loyalties that reemerged full-blown in the 1990s as all the great ideologies, save democracy, collapsed worldwide—and even democracy was none too stable outside the West.[133]

130 Herman O. Stekler, *The Structure and Performance of the Aerospace Industry* (Berkeley: University of California Press, 1965), pp. 14, 34; Roger E. Bilstein, *The American Aerospace Industry* (New York: Twayne Publishers, 1996), pp. 226–227; Richard Lampl, exec. ed., *The Aviation & Aerospace Almanac* (New York: McGraw-Hill, 1996), p. 703.

131 On this subject, see R. Earl McClendon, *Autonomy of the Air Arm* (Washington, DC: Air Force History and Museums Program, 1996); Bernard C. Nalty, ed., *Winged Shield, Winged Sword: A History of the USAF* (Washington, DC: Air Force History and Museums Program, 1997); Herman S. Wolk, *Toward Independence: The Emergence of the United States Air Force, 1945–1947* (Washington, DC: Air Force History and Museums Program, 1996).

132 See Roger D. Launius, "End of a Forty Year War: Demobilization in the West Coast Aerospace Industry After the Cold War," *Journal of the West* 36 (July 1997): 85–96.

133 On the reorientation of world politics in the 1990s, see John Lewis Gaddis, "Toward the Post-Cold War World," *Foreign Affairs* 70 (spring 1991): 101–114; Judith Goldstein and Robert O. Keohane, ed., *Ideas and Foreign Policy: Beliefs, Institutions, and Political Change* (Ithaca, NY: Cornell University Press, 1993); Francis Fucayama, "The End of History," *The National Interest* 16 (summer 1989): 3–18; Max Singer and Aaron Wildavsky, *The Real World Order: Zones of Peace, Zones of Turmoil* (Chatham, NJ: Chatham House, 1993); James M. Goldgeier and Michael McFaul, "A Tale of Two Worlds: Core and Periphery in the Post–Cold War Era," *International Organization* 46 (spring 1992): 467–491; Kenneth N. Waltz, "The Emerging Structure of International Politics," *International Security* 18 (fall 1993): 44–79; Zbigniew Brzezinski, *Out of Control: Global Turmoil on the Eve of the Twenty-first Century* (New York: Scribner, 1993); Daniel Patrick Moynihan, *Pandemonium: Ethnicity in International Politics* (New York: Oxford University Press, 1993); William S. Lind, "North-South Relations: Returning to a World of Cultures in Conflict," *Current World Leaders* 35 (December 1993): 1073–1080; Donald J. Puchala, "The History of the Future of International Relations," *Ethics and International Affairs* 8 (1994): 177–202; Samuel P. Huntington, *The Clash of Civilizations and the Remaking of World Order* (New York: Simon and Schuster, 1997).

In the two-plus decades since the fall of the Soviet Union, the U.S. military has placed much faith in bombing, both as a quick attack tactic and as a pre-invasion one. What the future of airpower may bring is anyone's guess, but the rise of airpower technology available for the United States is a significant determinant of the structure and possibilities of U.S. military force applications. At the same time, the rise of U.S. military aerospace capability has not significantly changed the nature of warfare. As analyst Anthony H. Cordesman observes: "War still requires a presence on the ground and a willingness to take casualties. This lesson should be obvious. Air power is incapable of holding territory, dealing with political issues, gathering human intelligence, and destroying dug-in enemy positions. It cannot seal off territory or deal with highly dispersed forces."[134] Ironically, this means that wars may only be won by putting forces on the ground, from which casualties will result. For all of the apparent changes, easy, casualty-free victory is no more possible today, even with the enormous aerospace power available to the United States, than it was during the first use of airplanes in combat. Still, the government's involvement in this arena is enormous and will remain so indefinitely.

The investment in military aeronautical technology indirectly affected the course of commercial aviation, whereas other activities directly benefited the construction of the air transportation system in the United States. Well into the 20th century, there was in the United States little appreciation of scientific and technical research and even less inclination to allocate government funding for such an uncertain activity. But because of a truly poor response at the time of World War I, the U.S. created the National Advisory Committee for Aeronautics (NACA) in 1915. Sentiment for some sort of center of aeronautical research had been building for several years. At the inaugural meeting of the American Aeronautical Society in 1911, its members discussed a national laboratory financed by the public treasury. But the American Aeronautical Society's dreams were frustrated by bureaucratic infighting and questions about the appropriateness of government investment in technological R&D. Only through the passage of enabling legislation for the NACA on 3 March 1915, as a rider to the Naval Appropriations Act, did anything happen. In this legislation, Congress established the NACA "to supervise and direct the scientific study of the problems of flight, with a view to their practical solution."[135]

The NACA became an enormously important government research and development organization for the next half century, materially enhancing the development of aeronautics. It pursued investigations that promised the compilation of fundamental aeronautical knowledge applicable to all flight, rather than working on a specific type of aircraft design, to avoid any charge of catering to a particular aeronautical firm. Most NACA research was accomplished "in-house" by scientists or engineers on the federal payroll. The results of these activities appeared in more than 16,000 research reports of one type or another, distributed widely for the benefit of all. As a result of this work, the NACA received the coveted Robert J. Collier Trophy, given annually for "great" achievement in aeronautics and astronautics in America, five times between 1929 and 1954.[136]

---

134 Anthony H. Cordesman, "The Old-New Lessons of Afghanistan," 4 March 2002, Center for Strategic and International Studies, Washington, DC.

135 The story of the NACA's creation is told in Roland, *Model Research*, 1:1–25; Roger E. Bilstein, *Orders of Magnitude: A History of the NACA and NASA* (Washington, DC: NASA SP-4404, 1989), chapter 1.

136 On this subject, see Pamela E. Mack, ed., *From Engineering Science to Big Science: The NACA and NASA Collier Trophy Research Project Winners* (Washington, DC: NASA SP-4219, 1998).

The NACA's research was conducted in government facilities, and its government scientists and engineers developed a strong technical competence, a commitment to collegial in-house research conducive to engineering innovation, and a definite apolitical perspective. The 8,000 employees worked in locations including a small Washington headquarters; three major research laboratories—the Langley Aeronautical Laboratory established in Hampton, Virginia, in 1917, the Ames Aeronautical Laboratory activated near San Francisco in 1939, and the Lewis Flight Propulsion Laboratory built in Cleveland, Ohio, in 1940—and two small test facilities in Muroc Dry Lake in the high desert of California and Wallops Island, Virginia. This organization remained a significant entity until it was transformed into NASA in 1958. The National Air and Space Act of 1958 gave NASA a broad mandate to "plan, direct, and conduct aeronautical and space activities"; to involve the nation's scientific community in these activities; and to disseminate widely information about them.[137]

The vast majority of breakthroughs in aeronautical technology have been the result of R&D organizations usually funded by government largesse. The experience of the NACA and NASA suggests that great leaps forward in technological capability almost always require significant long-term investment in research and development—research and development that does not have explicit short-term return to the "bottom line" and may not yield even long-term economic return. Without that large-scale investment in aerospace technology, however, the United States will become a second-class aerospace power. I would suggest that today we are on the road to becoming one. That is the result of two related developments. First, since the end of the Cold War and the ensuing belief that the United States stood alone as the world's only superpower, the level of aerospace R&D investment by the federal government has eroded. It was no longer viewed as necessary for national defense. Second, many public officials believe—mistaken though they are—that aerospace technology is mature and that private industry should be able to sustain aerospace advances without significant government investment.

Another major theme in government involvement in aerospace activities is in the regulation of aerospace commerce. The earliest involvement came with legislation to manage an airmail system, but in 1926, the Air Commerce Act assigned responsibility for the fostering of air commerce to the Commerce Department. With this went responsibility to establish airways, test and license pilots, inspect and certificate aircraft, establish navigation systems, investigate accidents, and generally provide for an orderly development of American aviation. Bumps and bruises were inflicted on all sides during the years that followed as a new technology and a new industry, as well as a relatively new government regulatory thrust, began to be played out. By the time of the creation of the Civil Aeronautics Authority in 1938, however, the major trends had been developed and much that followed was refinement.[138]

Three additional pieces of legislation were important in the regulation of aerospace operations. The Federal Aviation Act of 1958 transformed the Civil Aeronautics Authority into the Federal Aviation Agency (FAA), giving it broad powers to manage and regulate commercial aviation in the United States.

---

137 On the creation of NASA, see Roger D. Launius, *NASA: A History of the Civil Space Program* (Malabar, FL: Krieger Pub. Co., 1994), pp. 29–42.

138 See Arnold E. Briddon and Ellmore A. Champie, *Federal Aviation Agency Historical Fact Book: A Chronology, 1926–1963* (Washington, DC: Federal Aviation Agency, 1966).

In 1966, the FAA underwent a minor name change—Agency to Administration—and was assigned as a major component of the newly established Department of Transportation.

Perhaps the most important regulatory action since the Air Commerce Act of 1938, however, was the Airline Deregulation Act of 1978. In this legislation, Congress ended the federal enforcement of route structures and prices and allowed competition to reign in flights between American cities. This broke the near-monopolies of the major carriers—especially American, TWA, United, Eastern, Western, Northwest, Delta, and Braniff Airlines—and opened the door for numerous smaller and nonscheduled air carriers to enter the passenger market. Several of the major airlines either went out of business—such as Braniff and Eastern—and others, such as Delta Airlines, absorbed numerous other firms weakened by the competition. The airline business, therefore, became a sharply different arena than it had been before 1978. Whether the Airline Deregulation Act has ultimately been good for the nation and its flyers remains an open and hotly contested question.[139]

Among the key issues to be explored are the origins and development of airports in the United States and their governance and polity. Across the country, Americans take for granted the convenience of air flight from one city to another. The federal role in managing air traffic and the cooperative corporate planning of major airlines mask, to some degree, the fact that those airports are local public responsibilities that work in concert with federal and private interests. The early history of experimentation and innovation in the development of municipal airports—beginning with pressures from the U.S. Post Office and the military, neither of which had the independent resources to develop a network of terminals—led to American cities becoming responsible for air access.

Later, the federal government provided assistance for airport construction, maintenance, and improvement, especially through the New Deal Works Progress Administration (WPA) program. As part of a general trend during the 1930s toward a strong, direct relationship between cities and the federal government, cities began to lobby for federal aid for their airports, a demand that was eventually met when World War II increased the federal stakes in their functioning. The evolution continued into the era of the Cold War, and a unique public-private relationship has remained the norm to the present (see figure 5).[140]

139 A beginning in this direction may be found in Roger E. Bilstein, *Flight Patterns: Trends of Aeronautical Development in the United States, 1918–1929* (Athens: University of Georgia Press, 1983); Nick A. Komons, *Bonfires to Beacons: Federal Civil Aviation Policy Under the Air Commerce Act, 1926–1938* (Washington, DC: Smithsonian Institution Press, 1989); Janet R. Daly Bednarek, *America's Airports: Airfield Development, 1918–1947* (College Station: Texas A&M University Press, 2001).

140 See Janet R. Daly Bednarek, *America's Airports: Airfield Development, 1918–1947* (College Station: Texas A&M University Press, 2001).

Figure 5

Major U.S. Legislation Affecting Aerospace Activities

Public Act	Date Signed	Public Law (P.L.) Number	Major Provisions
Naval Appropriations Act of 1915	12 April 1915	63-271	A rider to this appropriations act established the National Advisory Committee for Aeronautics "to supervise and direct the scientific study of the problems of flight, with a view to their practical solution."
Air Mail Act of 1925	2 February 1925	69-309	The Kelly Act authorized the Post Office Department to contract for the delivery of domestic mail by commercial air carriers.
Air Commerce Act of 1926	20 May 1926	69-254	Instructed the Secretary of Commerce to foster air commerce, designate and establish airways, set up air navigation aids, arrange for R&D, license pilots, inspect and certificate aircraft, and investigate accidents.
Army Air Corps Act of 1926	2 July 1926	69-446	Renamed the Air Service as the Army Air Corps and provided for an Assistant Secretary of War for Air and for a five-year Air Corps expansion program.
Air Mail Act of 1930	29 April 1930	71-178	The Watres Act amended the 1925 Air Mail Act to give the Postmaster General broad regulatory control over route locations, route consolidations and extensions, contract bidding conditions, service conditions, equipment and personnel, and compensation.
Air Mail Act of 1934	12 June 1934	73-308	The Black-McKellar Act provided for the commercial contracting of airmail routes throughout the United States. Also created the Federal Aviation Committee to set broad policy on all phases of aviation and the relation of the government to it.
Civil Aeronautics Act of 1938	23 July 1938	75-706	Created the Civil Aeronautics Authority and Air Safety Board, both with broad powers to establish and operate airways and to regulate commercial air operations.

Public Act	Date Signed	Public Law (P.L.) Number	Major Provisions
National Defense Act of 1940	3 April 1939	76-18	Authorized the Army Air Corps to develop and procure 6,000 new airplanes, to increase personnel to 3,203 officers and 45,000 enlisted, and appropriate $300 million.
Civilian Pilot Training Act of 1939	27 June 1939	76-153	Established the Civilian Pilot Training Program under the management of the Civil Aeronautics Authority to train pilots at various educational institutions in the United States as a war preparedness measure.
Federal Airport Act of 1946	13 May 1946	79-377	Appropriated $500 million for continental United States and $20 million for Alaska and Hawaii for the construction of airports on a matching fund basis.
National Security Act of 1947	26 July 1947	80-253	Abolished the Departments of War and the Navy and established in their place the Department of Defense. In so doing, split the Army Air Forces out and made it a separate service, the United States Air Force, coequal with the U.S. Army and the U.S. Navy.
Airways Modernization Act	14 August 1957	85-133	Established the Airways Modernization Board "to provide for the development and modernization of the national system of navigation and traffic control facilities to serve present and future needs of civil and military aviation."
National Air and Space Act of 1958	29 July 1958	85-568	Transformed the National Advisory Committee for Aeronautics into the National Aeronautics and Space Administration.
Defense Reorganization Act of 1958	6 August 1958	85-599	Provided for a stronger Secretary of Defense, vesting control of research and development activities in that office. This provided for more centralized management of aerospace R&D.
Federal Aviation Act	23 August 1958	85-726	Transformed the Civil Aeronautics of 1958 Authority into the Federal Aviation Agency and made the Civil Aeronautics Board an independent organization. The new organization had broad powers to manage and regulate commercial aviation in the United States.

Public Act	Date Signed	Public Law (P.L.) Number	Major Provisions
"Crimes in the Sky" Act	5 September 1961	87-197	Amendment to Federal Aviation Act of 1958 to provide for the enforcement of crimes committed in the air, especially for interfering with the performance of duties by flight crews.
Communications Satellite Act of 1962	31 August 1962	87-624	Created Communication Satellite Corporation, a public-private corp. managing satellite communications for the United States.
Department of Transportation Act of 1966	15 October 1966	89-670	Created the Department of Transportation as a cabinet-level organization. The Federal Aviation Agency was assigned to the new department and given the name Federal Aviation Administration.
National Science and Technology Policy, Organization, and Priorities Act of 1976	11 May 1976	94-282	Created an Office of Science and Technology Policy reporting to the President.
Airline Deregulation Act	24 November 1978	95-504	Provided for fare reductions of up to 70 percent without Civil Aeronautics Board (CAB) approval and the immediate entry of air carriers into routes not protected by other carriers. The CAB's regulation of fares, routes, and mergers would be phased out by 1983, and unless Congress acted, the CAB would shut down by 1985.
Land Remote Sensing Commercialization Act	17 July 1984	98-365	Commercialized the Landsat remote sensing system launched by the United States in the 1970s.
Commercial Space Launch Act of 1984	30 October 1984	98-575	Commercialized launch operations within the United States to open them to competition.
Department of Defense Reorganization Act of 1986	1 October 1986	99-433	The Goldwater-Nichols Act provided for greater control of the individual services at the Secretary level and centralized even more aerospace R&D activities.

Sources: U.S. Senate, Committee on Commerce, Science, and Transportation, *Space Law and Related Documents: International Space Law Documents, U.S. Space Law Documents* (Washington, DC: Government Printing Office, 1990), pp. 443–605; Alex Roland, *Model Research: The National Advisory Committee for Aeronautics, 1915–1958* (Washington, DC: National Aeronautics and Space Administration, 1985), 2:393–422; Arnold E. Briddon and Ellmore A. Champie, *Federal Aviation Agency Historical Fact Book: A Chronology, 1926–1963* (Washington, DC: Federal Aviation Agency, 1966); Richard I. Wolf, ed., *The United States Air Force: Basic Documents on Roles and Missions* (Washington, DC: Office of Air Force History, 1987), pp. 325–338.

The rapid growth in aerospace technology within the United States during the 20th century was in part a result of investment by the United States government. Since the end of the Cold War, that investment has withered, and as it did so national leadership in the field declined as well. Between the 1960s and the 1990s, the share of the market enjoyed by American aerospace manufacturers fell sharply as foreign corporations—both private and state-run—gained greater portions of the market. In 1986, for example, United States high-technology imports exceeded exports for the first time. The aerospace industry was one of the only remaining fields with a trade surplus, 90 percent of which was from sales of aircraft and aircraft parts. Compared to an overall U.S. trade deficit in manufactured goods of $136 billion in 1986, the aerospace industry had a surplus of $11.8 billion. But the U.S. lead in aerospace was shrinking rapidly. In 1980, the U.S. market share of large civil transport sales was 90 percent. By 1992, that percentage had dropped to 70 percent and was in danger of falling even further. The lead in the commuter aircraft market had already been lost. During the 1990s, the United States lost its lead in the space launch market as well. Several factors account for this loss of market share and are complex.

First, there are the inherent difficulties of the aerospace marketplace. As aerospace technology became more complex and expensive, it also became more difficult for individual companies to shoulder the entire financial burden for researching and developing new technology and products themselves. Building airplanes has always been a marginal economic enterprise in all its myriad permutations. Aerospace manufacturers literally bet the company on a new design because of the enormous cost associated with developing an aircraft or rocket. Malcolm Stamper, former president of Boeing Aircraft Corporation, remarked that "Locating the break-even point is like finding a will-o'-the-wisp."[141] Not until 20 to 35 production aircraft have actually been manufactured do production costs become predictable. For rockets and other space technology, which do not have large production runs, the economics of manufacturing are even more problematic.[142]

Second, American aerospace executives were too often complacent in maintaining their competitive technological edge. Aerospace corporations, like a lot of other organizations, have a decided "not invented here" syndrome. Ideas emanating from beyond the recognized corporate structure too often get short shrift: to cite but two examples, Northrop Aircraft Corporation's hesitation to embrace retractable landing gear in the 1920s and Boeing's rejection of the so-called "glass cockpit" technology in the 1980s. While the "glass cockpit" offered cutting-edge avionics displays, this American-made technology found its first use at Airbus Industrie in Europe. Airbus made it a centerpiece of its newest generation of transports, in the process helping itself compete more effectively in the marketplace. Losing market share as a result, Boeing raced to adopt the new technology into its own designs.[143]

Third, there has been the success of industrial policy by the nations of Europe aimed at securing greater market share for non-U.S. aerospace companies. These governments often directly subsidize their national

---

141 Quoted in John Newhouse, *The Sporty Game* (New York: Alfred A. Knopf, 1982), p. 4.

142 Senate Committee on Armed Services, *Hearings, Weapons Systems Acquisition Process* (Washington, DC: Government Printing Office, 1972), 92d Cong., 1st Sess., p. 152.

143 See Walter G. Vincenti, "The Retractable Airplane Landing Gear and the Northrop 'Anomaly': Variation-Selection and the Shaping of Technology," *Technology & Culture* 35 (January 1994): 1–33; Lane E. Wallace, *Airborne Trailblazer: Two Decades with NASA Langley's Boeing 737 Flying Laboratory* (Washington, DC: NASA SP-4216, 1994), pp. 26–39.

manufacturers. There is no question that one of the major reasons for the European community to invest in aerospace technology has been to wrest economic market share from the United States. The Europeans have developed an industrial policy aimed at this goal, and they have been quite successful. Less successful, but nonetheless making inroads, are the Japanese, who have long pursued policies, and directly subsidized key industries, to help move the fruits of basic research to the marketplace for the purpose of gaining economic advantage vis-à-vis the United States.[144]

Finally, a major problem of the aircraft business was its cyclic nature, leading to boom and bust periods. Complicated by the enormous infrastructure necessary to support the design and manufacture of aircraft, these firms were exceptionally limited as to their markets and their capabilities. President Ronald Reagan's science advisor noted in 1982 that "aircraft are now the dominant common carrier for inter-city travel, and the safety and control of that travel are a federal responsibility."[145] He recommended pressing hard for government support of basic research that could then be transferred to American private firms.

Since 1903, the United States has spent more than 1 trillion dollars (in 2013 dollars) on developing aerospace technology, on the management of the infrastructure necessary to support its operations, and on the military and other practical applications that it affords. Accordingly, through a century of heavier-than-air flight, the federal government has been the major actor in developing and using aircraft. The United States did not have to make that investment—it could have chosen to act like many other major nations such as China and Brazil and Turkey—but because it did, the nation became the foremost air- and spacefaring power in the world. In truth, the impact of federal investments in aviation and space has been felt most directly in aerospace science and engineering, but it has also rippled through other fields of scientific and technological endeavor and across social, cultural, economic, and political arenas in the United States and worldwide. Apart from past theoretical developments in physics and the art and science of flight, federally supported R&D efforts have been a major factor—in some instances, the *key* factor—driving innovation.

The U.S. government's investment in both aviation and space has supported activities performed by a wide range of parties, including both civilian and military federal agencies, nonprofit quasi-governmental entities, private corporations, and educational institutions. A number of important conclusions emerge from this review of federal aerospace investment.

First, for good or otherwise, critical innovation in both the space and aviation sectors has been driven by external crises. Both world wars, particularly World War II, had a motivating effect on aerospace engineering and mass-production processes. Industrial techniques developed by Henry Ford for automobile production were applied to the war effort—Ford Motor Company mass-produced bombers and other aircraft in

144 See, for example, the Convention for the Establishment of a European Space Agency (CSE.CD(73)19. rev.7: Paris, 30 May 1975). Article VII (l) (b) states: "The industrial policy which the Agency is to elaborate and apply by virtue of Article II (d) shall be designed in particular to:...b) improve the world-wide competitiveness of European industry by maintaining and developing space technology and by encouraging the rationalisation and development of an industrial structure appropriate to market requirements, making use in the first place of the existing industrial potential of all Member States."

145 Bowie, *University-Business Partnerships*, p. 19.

numbers unimaginable before the war.[146] Following World War II, the Cold War saw the development of intercontinental ballistic missiles that provided the technological foundation and legacy for space launch vehicles.[147] This era also witnessed the development of advanced reconnaissance programs, such as the SR-71 aircraft and the Corona satellite, to gather information about areas that were otherwise inaccessible. These activities led directly to many of the present space science efforts, such as the Hubble Space Telescope, whose technologies and advanced engineering were based on these early intelligence programs.[148] The launch of Sputnik and piloted Russian spacecraft and the outcry that followed so stung the United States that it was propelled to accelerate the U.S. human spaceflight programs to a vast scale and cost previously considered unacceptable by the public.[149]

Second, the federal government's role in fostering innovation appears to have followed different paths in aviation R&D. Specifically, in aviation, the government, from virtually the beginning, focused on supporting inquiries into basic principles of aerodynamics and materials and propulsion, as well as creating facilities for conducting basic aviation science with the results available to all. This investment, in turn, helped the private aviation industry to progress from a small-scale, essentially craft-centered node of development and production to the vast, mass-production enterprise it became, with its own flourishing R&D component. By contrast, the federal government's role in fostering R&D in space

146 See Jacob Vander Meulen, *The Politics of Aircraft: Building an American Military Industry* (Lawrence: University Press of Kansas, 1991); John B. Rae, *Climb to Greatness: The American Aircraft Industry, 1920–1960* (Cambridge, MA: MIT Press, 1968); I. B. Holley, Jr., *Buying Aircraft: Material Procurement for the Army Air Forces* (Washington, DC: Center for Military History, 1964); Roger D. Launius, "World War II Military Aviation in the Rockies: From Natural to National Resource," *Journal of the West* 32 (April 1993): 86–93.

147 See Ray A. Williamson and Roger D. Launius, "Rocketry and the Origins of Spaceflight," in Roger D. Launius and Dennis R. Jenkins, eds., *To Reach the High Frontier: A History of U.S. Launch Vehicles* (Lexington: University Press of Kentucky, 2002), pp. 33–69; Launius, "Between a Rocket and a Hard Place," in Lambright, ed., *Space Policy in the 21st Century*, pp. 15–54.

148 Three important recent books on the early satellite reconnaissance program have been published: Dwayne A. Day, John M. Logsdon, and Brian Latell, eds., *Eye in the Sky: The Story of the Corona Spy Satellite* (Washington, DC: Smithsonian Institution Press, 1998); Robert A. McDonald, *Corona Between the Sun and the Earth: The First NRO Reconnaissance Eye in Space* (Bethesda, MD: ASPRS Publications, 1997); Curtis Peebles, *The Corona Project: America's First Spy Satellites* (Annapolis, MD: Naval Institute Press, 1997). See also William E. Burrows, *Deep Black: Space Espionage and National Security* (New York: Random House, 1987); Jeffrey T. Richelson, *America's Secret Eyes in Space: The U.S. Keyhole Spy Satellite Program* (New York: Harper and Row, 1990).

149 This argument has been effectively made in Rip Bulkeley, *The Sputniks Crisis and Early United States Space Policy: A Critique of the Historiography of Space* (Bloomington: Indiana University Press, 1991); Robert A. Divine, *The Sputnik Challenge: Eisenhower's Response to the Soviet Satellite* (New York: Oxford University Press, 1993); Dwayne A. Day, "New Revelations About the American Satellite Programme Before Sputnik," *Spaceflight* 36 (November 1994): 372–373; R. Cargill Hall, "Origins of U.S. Space Policy: Eisenhower, Open Skies, and Freedom of Space," in John M. Logsdon, gen. ed., *Exploring the Unknown: Selected Documents in the History of the U.S. Civil Space Program, Volume I, Organizing for Exploration* (Washington, DC: NASA SP-4407, 1995), pp. 213–229; Roger D. Launius, "Eisenhower, Sputnik, and the Creation of NASA: Technological Elites and the Public Policy Agenda," *Prologue: Quarterly of the National Archives and Records Administration* 28 (summer 1996): 127–143; R. Cargill Hall, "Earth Satellites: A First Look by the United States Navy," in R. Cargill Hall, ed., *Essays on the History of Rocketry and Astronautics: Proceedings of the Third Through the Sixth History Symposia of the International Academy of Astronautics* (San Diego, CA: Univelt, Inc., 1986), pp. 253–278; R. Cargill Hall, "The Eisenhower Administration and the Cold War: Framing American Aeronautics To Serve National Security," *Prologue: The Journal of the National Archives* 27 (spring 1995): 61–70; Dwayne A. Day, "Not So Black and White...: The Military and the Hubble Space Telescope," *Space Times: Magazine of the American Astronautical Society* 34 (March–April 1995): 20–21.

innovation often appears to have reversed the typical progression from basic research to development. Instead, federal investment in space innovation concentrated on technology development and engineering, with little emphasis on basic research. By the time the U.S. space program began scaling up to produce spacecraft in the 1950s, the basic principles of rocketry and spaceflight were already well understood. The practical problems of launching viable spacecraft, however, were not. In turn, spaceflight facilitated the deployment of new scientific instrumentation and the conduct of new experiments that fueled advances in basic science.[150]

Third, the cross-fertilization between national security (defined as the military and intelligence communities) and civilian innovation has been central to the success of both the aviation and space enterprises. Because the military and civilian industrial bases in both sectors overlap substantially and many technologies are "dual-use," U.S. national security and civilian programs have often benefited from innovations emerging from the other domain or from joint efforts. Yet different organizational missions have also led to differing institutional focus and emphasis: for the NACA and NASA, on basic aviation and space science and R&D; for the national security community, including the military services, more often applied R&D of advanced weapon systems. While the development of a particular technology by one agency may not have led to the direct production of an operational system by another agency and thus could be construed as a failure of innovation, in other instances that technology development led to unanticipated breakthroughs in other technologies or system approaches that could be applied across organizational and institutional boundaries. Applying performance measures to pure research has not always worked, yet many would argue that taxpayer dollars for focused federal investment in aerospace R&D should provide a worthwhile return on investment. Nevertheless, opportunities for cross-fertilization between federal organizations, driven by decreasing federal investments in aerospace R&D and other factors, are having the effect of encouraging greater numbers of joint or interagency activities to pursue dual-use technologies.

Fourth, the evolution of large-scale engineering techniques and methodologies has yielded important lessons for similar-scale projects in other areas. These accumulated lessons have enduring relevance—from the days of the NACA and large-scale wind tunnels for testing aircraft to the large-scale engineering projects involved in human spaceflight. From the Mercury, Gemini, and Apollo programs to the Space Shuttle, the International Space Station, and the exploration of the Moon and Mars, they continue to guide and inform.[151] One can clearly speculate where the future of flight may go, but whatever the course, success will hinge on continued federal stimulation of private-sector activity so that in the future it can build on the foundation of past successes and failures.[152]

Between 1915 and the 1970s, government officials in the United States undertook a series of critical initiatives designed to create a commercial airline industry in private hands. Washington lawmakers saw the necessity of fostering new technology for the purposes of national security, economic competitiveness,

150 See Roger D. Launius, ed., *Innovation and the Development of Flight* (College Station: Texas A&M University Press, 1999).

151 James E. Webb, *Space Age Management: The Large Scale Approach* (New York: McGraw-Hill Book Co., 1969), p. 15.

152 On the future of flight, see Paul McCready, "Atmospheric Talents," *Bulletin of the American Meteorological Society* 76 (June 1995): 1019–1021; Roger D. Launius and Howard E. McCurdy, *Imagining Space: Achievements, Predictions, Possibilities, 1950–2050* (San Francisco: Chronicle Books, 2001), pp. 166–168.

and pride and prestige. That latter reason was in no small measure, because although Americans had invented the airplane in 1903, by 1914 leadership in aviation technology had moved to Europe and the United States had been left in the dust. Catching up became an important driver for federal investment. Government organizations took a multifaceted approach: military investment; research and development; regulatory efforts aimed at promoting safety, efficiency, and expanded operations; and direct subsidies to commercial entities until the 1960s. Congress could have established a national airline run by civil servants, but instead created a favorable climate for private investment in airlines. For instance, the U.S. Congress established the NACA in 1915 to conduct research on flight, and in 1921, New York and New Jersey created a port authority with the power to issue bonds and collect fees for airfields.

In terms of space transportation, there are several lessons to be drawn from the aviation experience. Like the NACA, government agencies could conduct basic research and transfer that knowledge to private firms. In addition, NASA could transfer its operational responsibility to private carriers. Congress could also create the authority—modeled on various earlier efforts such as the Overseas Private Investment Corporation—to provide loans and insurance to space line firms. Either the U.S. government or states could establish spaceport authorities to manage operations from the ground to orbit; federal agencies could also regulate routes and fares. Many of these efforts are already under way, and we are on the verge of seeing a new age of entrepreneurial space transportation efforts. There are, however, challenges to this approach, not the least of which is that NASA has a critical path with specific milestone deadlines and is hesitant to change this approach; the loans/insurance incentives may not produce services in time; and liability issues are especially burdensome. Nonetheless, major steps have been taken toward this capability in the last decade.

Creating the Telephone Industry

Everyone agrees—or almost everyone, since there are a few pretenders to this title and each has champions—that Alexander Graham Bell invented the telephone in 1876 in Boston, Massachusetts. A year later, he organized the Bell Patent Association to lease equipment to users.[153] At first, as the inventor of the telephone, Bell held all rights to the technology; as time progressed, he worked to ensure that it gained as near monopoly status as possible. What became known as the American Telephone and Telegraph Company expanded over time, operating the full system, to which subscribers paid a fee for use as well as equipment rental, built at its laboratory, Western Electric Company.[154]

At this point, the Bell system was essentially a Massachusetts corporation providing this newly developed technology to users in New England. The Massachusetts legislature assisted in its expansion by

153 Alexander Graham Bell, "Alexander Graham Bell Recalls Early Days of the Telephone," *Boston Daily Globe*, 5 November 1911; Nandini Chandar and Paul J. Miranti, "Integrating Accounting and Statistics: Forecasting, Budgeting, and Production Planning at the American Telephone and Telegraph Company During the 1920s," *Accounting & Business Research* 39, no. 4 (2009): 373–395.

154 "The Bell Telephone Company and the People," *New York Times*, 8 February 1886; Curtis Sweet, "The Passing of the Telephone Age," *Business Communications Review* 33, no. 4 (2003): 16–17; Louis Galambos, "High-Technology Industries," *Business History Review* 66, no. 1 (spring 1992): 95–126.

subscribing government resources to it, increasing its capital by $20,000 to $30,000.[155] This enabled the expansion of the system over so-called "long lines" to New York City and other major cities in the United States. By 1895, the Bell system had over 300,000 phones, but this was nothing compared to the system's expansion over the next 10 years, to 2,284,587 phones in 1905.[156]

Until this time, the Bell system had been loosely organized—to call it a system was a misnomer—but in the first part of the 20th century, AT&T's leadership pressed to create a tightly managed, universally interconnected telephone system under a single company with a single set of policies. It moved effectively to undertake horizontal integration of the industry, advancing services broadly throughout American society and reducing competition while it increased its market share. In theory, economies of scale achieved in this way would increase the efficiency of the industry, although decreased competition might not result in savings to the customer.[157]

AT&T began buying out the competition and consolidating various telephone providers into an ever-larger network. Company president Theodore Vail aggressively pursued this corporate strategy, epitomizing it by the slogan, "One Policy, One System, Universal Service." AT&T began buying out competing companies (including the powerful telegraph company, Western Union), reducing their prices to gain control of competitors and then, once establishing a monopoly, raising the prices charged. The Mann-Elkins Act of 1910 extended the jurisdiction of the Interstate Commerce Commission (ICC) to include the regulation of telephone companies, and these business practices immediately focused attention on AT&T.[158]

During World War I, the United States Post Office took control of all telegraph and telephone services for one year and then returned the telephone and telegraph services to private firms after the Armistice; at the end of the war, the Post Office returned these corporate assets to private control with the caveat that AT&T must accept close monitoring and tight regulation as a public trust under the ICC. The federal government and AT&T co-established terms of government regulation and ownership:

> The federal government...agreed to pay to AT&T 4½ percent of the gross operating revenues of the telephone companies as a service fee; to make provisions for depreciation and obsolescence at the high rate of 5.72 percent per plant; to make provision for the amortization of intangible capital; to disburse all interest and dividend requirements; and in addition, to keep the properties in as good a condition as before.[159]

155 "The Bell Telephone Company and the People," *New York Times*, 8 February 1886.

156 Galambos, "High-Technology Industries," pp. 95–126.

157 Jill Hills, "Regulation, Innovation and Market Structure in International Telecommunications: The Case of the 1956 TAT1 Submarine Cable," *Business History* 49, no. 6 (2007): 868–885; Glenn Porter, *The Rise of Big Business, 1860–1920* (Arlington Heights, IL: Harlan Davidson, 1992 ed.).

158 David Gabel, "Competition in a Network Industry: The Telephone Industry, 1894–1910," *Journal of Economic History* 54, no. 3 (September 1994): 543–572.

159 N. R. Danielian, quoted in Adam D. Thierer, "Unnatural Monopoly: Critical Moments in the Development of the Bell System Monopoly," *Cato Journal* 14, no. 2 (fall 1994): 267–285, quotation from p. 275.

Thereafter, AT&T had to negotiate with governmental entities for rates and standards of service. The company did very well in this heavily regulated environment. By 21 January 1919, long-distance rates had increased by 20 percent. Additionally, the company was also able to establish long-distance rates based on average cost, enabling them to charge more money to callers in big cities to subsidize callers in rural areas. Before the end of 1919, AT&T had received an estimated $42 million in rate increases. "Additionally, the government cut AT&T a $13 million dollar check at the end of the period to cover any losses they may have incurred, despite the fact that none were evident."[160] This was only the beginning. By 1925, almost every state had instituted strict rate regulations that discouraged telephone competition.

In 1934, the regulation of interstate telephone and telegraph services was transferred from the ICC to the Federal Communications Commission. The FCC was given the power to act in public interest over the telephone service, setting up AT&T as a government-sanctioned monopoly. Not without misgivings, elected officials embraced the AT&T monopoly. The argument for it rested on the concept of "natural monopoly," a workable approach when a single firm dominates the market and when economies of scale are necessary to ensure effective service. Telecommunications represented a classic case of one firm being viewed as capable of serving consumers at lower costs than two or more firms.[161] As economist Adam D. Thierer commented:

> For example, telephone service traditionally has required laying an extensive cable network, constructing numerous call switching stations, and creating a variety of support services, before service could actually be initiated. Obviously, with such high entry costs, new firms can find it difficult to gain a toehold in the industry. Those problems are compounded by the fact that once a single firm overcomes the initial costs, their average cost of doing business drops rapidly relative to newcomers.[162]

This proved a compelling argument for Congress: "There is nothing to be gained by local competition in the telephone business."[163] State and national regulatory agencies actively sought to reduce competition for AT&T during the interwar period, calling it "wasteful duplication."[164]

The Communications Act of 1934 transferred the control of telephone regulations from the ICC to the newly established Federal Communications Commission (FCC). The FCC regulated rate changes, consolidations, connections, and the licensing of new companies. Even so, AT&T continued to operate as a monopoly until the 1970s, when the FCC suspected that AT&T was violating antitrust law. In 1974, *United States vs. AT&T* was filed by the Department of Justice, and in 1982, a settlement was reached in which the Bell system divided into several separate businesses.[165]

---

160 Ibid., p. 276.

161 D. F. Spulber, "Deregulating Telecommunications," *Yale Journal of Regulation* 12, no. 1 (1995): 25–67.

162 Thierer, "Unnatural Monopoly," p. 268.

163 Quoted in G. H. Loeb, "The Communications Act Policy Toward Competition: A Failure To Communicate," *Duke Law Journal* 1 (1978): 1–56, quote from p. 14.

164 Gabel, "Competition in a Network Industry," p. 562.

165 Paul J. Miranti, "Integrating Accounting and Statistics: Forecasting, Budgeting and Production Planning at the American Telephone and Telegraph Company During the 1920s," *Accounting and Business Research* 39, no. 4 (2009): 373.

At the time of divestiture, the Bell System was composed of American Telephone & Telegraph, its 22 operating companies, Western Electric, and Bell Laboratories. This system served 84 percent of the nation's telephone subscribers while the other 16 percent was served by 2,100 independents. AT&T had a little over $5 billion in assets, 685,000 stockholders, and an annual income of slightly more than $1 billion. All of Bell Systems resources—24,000 buildings; 177,000 motor vehicles; 1,000,000 employees; 142,000,000 telephones; and 1,700,000,000,000 miles of cable, microwave radio, and satellite circuits—had to be evenly distributed through nine viable corporate operations. Throughout the history of the land telephone, AT&T and Bell systems have given the United States an underground system of wire placement, metallic circuits, switchboards, long-distance communications between distant cities, and, above all, transcontinental telephone lines.[166]

The service provided to the American customer, and the costs incurred for this service, prompted the breakup of AT&T and the creation of the so-called "Baby Bells" in the 1980s: Bell South, Bell Atlantic, Nynex, Ameritech, SBC, U.S. West Communications, and Pacific Telesis. Much of this was a positive development. The Bell conglomerate was less innovative, less committed to customer service, and more costly to consumers than other systems. This changed with the Bell cartel's breakup, at least for a time. By the latter part of the 1990s, especially with the advancing capabilities of telecommunications available to serve consumers, various telephone, cable TV, and other service providers began to merge, create joint ventures, and enlarge market share. Every time that happened, the services might become more transparent and convenient to the user, but the costs for their use increased. Many have questioned the beneficial nature of these more recent changes, and some of have advocated for a stricter regulatory environment. There is no resolution at present to this conundrum.[167]

This brings to the fore the issue of satellite telecommunications and the role of the U.S. government in helping to bring it about, facilitate it, and regulate it from the 1960s to the present. With the dawn of the Space Age, AT&T sought to extend its telecommunications monopoly into space by gaining approval to build its own communications satellites and operate as an approved monopoly. The Eisenhower administration had been warm to this approach, approving the development of Telstar, launched in 1962, as a corporate activity. At the same time, the Kennedy administration had a different approach to satellite communications. It sponsored the Communications Satellite Act of 1962. Key to this was the public-private COMSAT Corporation chartered to oversee the operation of communications satellites. Some free-market advocates believed then and since that the federal government's intervention in this arena was heavy-handed and, in some instances, punitive. AT&T was quite willing to develop this technology without government involvement; the company questioned why it should not have been allowed to do so. The approach that a public-private entity would oversee space communications worked for a number of years but has been

166 W. Brooke Tunstall, "The Breakup of the Bell System: A Case Study in Cultural Transformation," *California Management Review* 28, no. 2 (1986): 110–124; Leland L. Johnson, "Technological Advance and Market Structure in Domestic Telecommunications," *American Economic Review* 60, no. 2 (1970): 204–208.

167 Tunstall, "Breakup of the Bell System," pp. 120–124; Jerry A. Hausman, "Competition in Long-Distance and Telecommunications Equipment Markets: Effects of the MFJ," *Managerial and Decision Economics* 16, no. 4 (July–August 1995): 365–383.

superseded by a new business arrangement. There is still considerable regulation—especially for launch, orbital slots, and frequencies—but the result has been more open than anything since earlier eras.[168]

In terms of lessons learned for space commercialization, there are several issues to be addressed. The government's role in providing patents is a given, but the granting of monopoly status and chartering special corporations with protections of a wide nature did not serve to foster an effective system that led to further enhancement of business opportunities. Will the government encourage private entrepreneurs to construct, own, operate, and use lunar communications networks, Mars communications networks, and deep space networks? This is a fundamental challenge for the future: Recent experience (Iridium, GPS) suggests that the cost of establishing certain space communications networks exceeds likely revenues. In such an environment, is the public-private partnership, with both sides investing, the best way forward?

Following the invention of the telephone in 1876, the federal government could have owned and operated telephone service—it did so during World War I—or it could have allowed a totally open market. Instead, it established phone companies as regulated monopolies under the FCC, with monopolistic privileges only removed in 1980. In essence, the following structure emerged:

- The government provided patents, granted monopoly status, and chartered corporations.
- The U.S. Attorney General allowed AT&T to control telephone service as a regulated monopoly (1913).
- AT&T established Bell Laboratories (1925); Bell Labs developed the first orbiting communications satellite (Telstar 1, 1962).
- Congress created COMSAT, a public-private corporation with monopoly status, to promote satellite communications (1962).
- COMSAT represented the United States in the formation of INTELSAT and became its managing company.

Might the U.S. government foster a private space communications system that can serve the needs of all users on a commercial basis, rather than having NASA own its on TDRSS satellites? What is the future of space communications?

Supporting Scientific Research in Antarctica

The American involvement in Antarctica dates to the 19th century, but it was limited to separate spheres of private commercial ventures and public scientific pursuits much of the time. Many of the 19th-century private interests in Antarctica centered on the sealing and whaling industries. By the early 20th century, according to Richard E. Byrd, "in summertime, its waters swarm with Norwegian whalers who annually harvest a revenue of $15,000,000 from their catch."[169] The expeditions of Robert F. Scott, Ernest Shackleton, and others revealed Antarctic deposits of iron, coal, titanium, and copper.

---

168 David J. Whalen, *The Origin of Satellite Communication* (Washington, DC: Smithsonian Institution Press, 2002); Andrew J. Butrica, ed., *Beyond the Ionosphere: Fifty Years of Satellite Communication* (Washington, DC: NASA SP-4217, 1997).

169 Richard Evelyn Byrd, *Little America: Aerial Exploration in the Antarctic The Flight to the South Pole* (New York: G.P. Putnam's Sons, 1930), p. 39.

Yet beyond the few natural resources discovered and the sealing and whaling industries, there was little private interest in the continent.

Richard Byrd recounted several reasons why this was the case in the interwar period. He noted that the continent was geographically isolated and its weather conditions were highly inhospitable. Prior to the heavy scientific investment of the postwar era, little was known about the continent in terms of how to navigate it, and little was known of its geology, geomagnetism, or glaciological formations. In this regard, there was little certainty of how navigational instruments would operate. Flight was still in its infancy and incapable of doing much to open the territory. Byrd fully believed that there were reasons to exploit the resources of Antarctica, but he recognized that use and profit had a long way to go before realization. He did not discount the scientific community's interest in Antarctica, but he wanted to advance commercial motivations.[170] Byrd recounted that "Antarctica[,] 'a vast wonderland laid out on a giant scale, in which littleness has no place,' cannot be judged, or appraised, according to limited values." He expressed disappointment: "[V]ainly did I [Richard Byrd] try to impress this fact upon a well-known American business man. 'But where's the money in it? Where's the profit[?]' he demanded."[171] After his failure to convince others of the commercial opportunities in Antarctica, scientific inquiry and discovery remained America's chief interests in Antarctica other than the offshore private whaling and sealing industry.

This did not change until the opening of major U.S. permanent sites on the continent during the International Geophysical Year (IGY) of 1957–58. The initial mission of the IGY was to study the geophysics of Earth's polar regions, but this mandate eventually expanded to include a wide range of scientific efforts. The Antarctic program fostered collaborative research with other countries and "comprises research by scientists selected from universities and other research institutions and operations and support by a contractor and other agencies of the U.S. Government."[172]

Everyone realized that human exploration of the polar regions and scientific research there were both costly and life-threatening. The U.S. federal government pursued these activities for the public good throughout much of the 20th century. These were fundamentally governmental activities. In particular, the government sought to establish human research stations in Antarctica, mobilizing considerable public resources to do so. The Navy and the National Science Foundation (NSF), the principal entities involved in Antarctic operations, led this effort, operating them for presumed altruistic scientific purposes.[173]

---

170 Lisle A. Rose, *Explorer: The Life of Richard E. Byrd* (Columbia: University of Missouri Press, 2006), pp. 405–411.

171 Byrd, *Little America*, p. 39.

172 National Science Foundation, "United States Antarctic Program," available online at *http://www.nsf.gov/geo/plr/antarct/usap.jsp*, accessed 13 June 2013.

173 Numerous works capture the American experience and motivation to explore the poles; see, in particular, Michael Robinson, *The Coldest Crucible: Arctic Exploration and American Culture* (Chicago: University of Chicago Press, 2006); Kenneth Bertrand, *Americans in Antarctica, 1775–1948* (Washington, DC: American Geographical Society, 1971); G. E. Fogg, *A History of Antarctic Science* (Cambridge, U.K.: Cambridge University Press, 1992); Stephen Pyne, *The Ice: A Journey to Antarctica* (New York: Ballantine Books, 1988). These are recent, ambitious works that address the various science missions undertaken in Antarctica and argue for their importance.

In many respects, the history of human actions in Antarctica mirrors the larger story of how the various great powers have interrelated since the conclusion of World War II. If one were to characterize it accurately throughout the last 50-plus years, the undeniable conclusion is that all parties have participated in an uneasy relationship in which they have recognized that they were better off cooperating to ensure that the icy continent was not developed. Instead, they sought to preserve it as a pristine continent. This approach to dealing with Antarctica is really a relic of the Cold War rivalries of the latter 1950s, and especially of the remarkable scientific endeavor known as the International Geophysical Year (IGY). Twelve nations participated directly in Antarctic research during that organized scientific research effort. The approach, in essence, ensured that since one side could not exploit the continent, no one would be allowed to do so.[174]

Because of the success of this approach during the IGY, which was managed by an international consortium, within a year afterward the 12 nations already in Antarctica met in Washington, DC, to sign the Antarctic Treaty, which "internationalized" Antarctica on a limited basis as a "continent dedicated to peace and science." Article IV of the Antarctic Treaty suspended (or "froze" in the official pun of the conference) all sovereignty claims to the continent for its duration, bringing to an end the active phase of very real disputes between Great Britain, Argentina, and Chile over control of the continent. To many people at the time, it appeared as if the idealism of science had trumped Cold War geopolitics. Historians have tended to follow this idealistic interpretation of the connection between the IGY and the Antarctic Treaty, and the southern continent tends to be held up as an all-too-rare example of scientific cooperation fostering political harmony.[175]

Of course, the IGY did indeed play an important role in the resolution of the Antarctic sovereignty dispute, but not in quite the idealistic way that the traditional narrative has suggested. The actual science of the IGY, and the improved understanding of the Antarctic environment that it facilitated, played an important role in the partial resolution of the question of sovereignty. As officials in the treaty nations, especially in Great Britain and the United States, learned more about the reality of the Antarctic environment through the work of the IGY—in particular the realization that it contained little or nothing of immediate economic value—they acceded to arguments in favor of internationalizing the continent. There was, in any eventuality, not much of a downside in the foreseeable future.[176]

---

174 See Roger D. Launius, James Rodger Fleming, and David H. DeVorkin, eds., *Globalizing Polar Science: Reconsidering the International Polar and Geophysical Years* (New York: Palgrave Macmillan, 2010).

175 M. J. Peterson, *Managing the Frozen South: The Creation and Evolution of the Antarctic Treaty System* (Berkeley: University of California Press, 1988); Jeffrey D. Myhre, *The Antarctic Treaty System: Politics, Law, and Diplomacy* (Boulder, CO: Westview Press, 1986); Steven J. Burton, "New Stresses on the Antarctic Treaty: Toward International Legal Institutions Governing Antarctic Resources," *Virginia Law Review* 65 (April 1979): 421–512; Paul Arthur Berkman, "Common Interests in the International Space of Antarctica," available online at Antarctic Treaty Summit 2009, *http://www.atsummit50.aq/media/journal-6.pdf*, accessed 21 January 2011.

176 The idealistic position is maintained in Richard S. Lewis, *A Continent for Science: The Antarctic Adventure* (New York: Viking Press, 1965); A. Elzinga, "Antarctica: The Construction of a Continent by and for Science," in Elisabeth Crawford et al., eds., *Denationalizing Science* (Dordrecht, The Netherlands: Kluwer Academic, 1993), 73–106. More realpolitik positions are offered in Peter Beck, *The International Politics of Antarctica* (London: Croom Helm, 1986); Klaus Dodds, *Geopolitics in Antarctica: Views from the Southern Oceanic Rim* (New York: John Wiley, 1997); Paul Arthur Berkman, *Science into Policy: Global Lessons from Antarctica* (San Diego, CA: Academic Press, 2002).

Accordingly, the United States led an effort to defuse geopolitical tensions in Antarctica by internationalizing the continent. As the various nations accepted this position, they found themselves members of the Antarctic Treaty System's (ATS's) "exclusive club," which continues to govern the continent to this day. Initially the Antarctic Treaty signatory countries disagreed on the question of the Soviet Union's role on the continent. U.S. officials, perhaps somewhat naively, believed that they could create a treaty regime for Antarctica that would exclude the Soviet Union. British officials—who were especially keen to resolve the dispute—argued, more realistically, that the communist superpower would have to be included for any internationalization of Antarctica to work. After some discussion, the British position prevailed.[177]

The Antarctic Treaty has been quite successful overall. In addition to having an intrinsic value of its own—especially at a time of growing awareness of the centrality of the southern continent to the global environment—science has also done much to keep the peace in Antarctica. Scientific cooperation has laid the basis for half a century of peaceful coexistence in a region that was becoming increasingly contentious in the 1940s and 1950s.[178]

Once a permanent presence in Antarctica was established, how would it be maintained? This would be, without question, a highly regulated environment because of the treaty system governing the Antarctic continent. Not only banning Great Power actions that might upset the status quo there, the treaty system also barred natural-resource exploitation. Despite attempts to deregulate and effectively open Antarctica to resource extraction, international policy has remained firmly opposed through the present. Accordingly, the Antarctic Treaty System has thus far kept corporate objectives in check. At the same time, it has turned off a profit motive for and expansion of access to the continent, in the process keeping prices of support high and fundamentally a government-supported activity.

With that as the case, there are a small number of organizations that could provide support to American efforts in Antarctica. Since there is no permanent population and no normal economic activity, supplies need to be transported in at government expense to sustain the American presence there. And it is important to note that the government presence is the only one in Antarctica because of the treaty system in place prohibiting commercial development. In the 1950s, the United States had traditionally been represented by the National Science Foundation (NSF) and the Navy. As late

177 Rip Bulkeley, "The Political Origins of the Antarctic Treaty," *Polar Record* 46 (2010): 9–11; Paul Arthur Berkman, "Common Interests in the International Space of Antarctica," *Polar Record* 46 (2010): 7–9; Christy Collis and Quentin Stevens, "Modern Colonialism in Antarctica: The Coldest Battle of the Cold War," in Gunter Lehman and David Nichols, eds., *Proceedings of the 7th Australasian Urban History/Planning History Conference* (Geelong, Australia: Deakin University, 2004), pp. 72–95, quote from p. 95. See also Quentin Stevens and Christy Collis, "Living in the Cold Light of Reason: Colonial Settlements in Antarctica," in Maryam Gusheh and Naomi Stead, eds., *Proceedings of Progress: The 20th Annual Conference of the Society of Architectural Historians, Australia and New Zealand* (2003): 291–297.

178 M. J. Peterson, "Antarctica: The Last Great Land Rush on Earth," *International Organization* 34, no. 3 (summer 1980); J. Dugger, "Exploiting Antarctica's Mineral Resources, *University of Miami Law Review* 33 (December 1978).

as 1982, President Ronald Reagan issued a Presidential Memorandum affirming that NSF would be the managing body for the United States' presence in Antarctica. It gave NSF responsibility to

- budget for and manage the entire United States national program in Antarctica, including logistic support activities so that the program may be managed as a single package;
- fund university research and federal agency programs related to Antarctica;
- draw upon logistic support capabilities of government agencies on a cost reimbursable basis; and
- use commercial support and management facilities where these are determined to be cost effective and will not, in the view of the [Antarctic Policy] Group, be detrimental to the national interest.[179]

It allowed other agencies to undertake short-term programs of scientific activity related to Antarctica, but only under strict regulation.

The American permanent presence really began with Operation Deep Freeze I (ODF), an expedition to Antarctica in 1955 and 1956 with the goal to establish a permanent settlement for the United States. An American force of 19 aircraft, 10 ships, and 1,000 men departed the United States for New Zealand, where the military already had in place icebreakers—the USS Glacier and the USS Edisto—and several transport planes at Christchurch International Airport. The IGY was the impetus for ODF, and the military was to provide logistics for the National Committee IGY scientists, conduct mapping and special projects, establish permanent stations as directed by the American scientific community (what would become NSF), and conduct scientific investigations for the Department of Defense (DOD).[180] ODF's first priorities were logistics and establishing the Little America IV station and what became Byrd Station. In order to begin air operations, ODF established an Air Operating Facility on the Ross Ice Shelf. This way, flights could be conducted on the Antarctic surface instead of relying on the constant presence of aircraft carriers.

The project relied on America's agreements with New Zealand to use the Christchurch International Airport as a reliable airport and byway for cargo. The Navy and Air Force carried out the operations during the Antarctic summer months between 1955 and 1956 to completion. Those included making nine successful long-range exploratory flights, building the Little America base, assembling the parts for Byrd Station, building the Air Operating Facility at McMurdo, exploring less-known parts of Antarctica in the Astrid Coast, and locating sites for the U.S.–New Zealand IGY station and IGY sites near the Weddell Sea and Vincennes Bay on the Knox Coast.[181]

---

179 "President's Memorandum Regarding Antarctica," Memorandum 6646, 5 February 1982, available online at *http://www.nsf.gov/geo/plr/ant/memo_6646.jsp*, accessed 15 September 2013.

180 "Deep Freeze I, Summary of Operations, 1955–56," Naval History & Heritage Command, available online at *http://www.history.navy.mil/ac/exploration/deepfreeze/ctf.htm*, accessed 23 July 2013.

181 Ibid.

Essentially, the first significant American initiative to establish a permanent human existence in Antarctica required a huge influx of government-funded personnel, materiel, and strong logistics support, which continues to this day. The Navy ended its involvement with the Antarctic Program in 1998. The Air Force and National Guard remain active players.

Human presence in Antarctica has increasingly included tourists since 1958, when 194 people made their way to the continent. The tourism industry began to flourish in the 1990s, when, for the first time, over 5,000 visitors reached Antarctica in 1992. When environmental protection became a significant pillar of the Antarctic Treaty System (ATS), tourism policy began to be stringently regulated. As tourism rose, so too did the regulations governing it. Signed by the ATS members at their 1991 meeting in Madrid, the Protocol on Environmental Protection and its seven annexes maintained Antarctica's status as a natural reserve for science and peace but further designated it as a continent free of mineral resource extraction (except in the interest of science), subjected human activity to environmental impact assessment (EIA), and ensured the protection of flora and fauna through pollution mitigation and waste disposal procedures. The Protocol also established a Committee for Environmental Protection (CEP) to advise and provide recommendations to the Antarctic Treaty Consultative Meeting (ATCM).[182]

One of the key regulatory instruments the Protocol called for was the EIA. The EIA is in place to assess numerous environmental concerns such as "translocating species whether floral or faunal," transitory damage from tourists, and chemical pollution; however, the EIA has some drawbacks.[183] Legal analysts Kees Bastmeijer and Ricardo Roura describe how the EIA assesses tourism's impact on Antarctica:

> The three tiered EIA system of the Protocol refers to whether impacts are less than, no more than, or more than "minor and transitory". The greater the intensity and duration of impacts, the greater the detail and scrutiny required in the EIA. Cumulative impacts, if and when they occur, may also be more serious. For example, tourism could plainly be a factor in introducing or translocating alien species or diseases, but this process is not well understood or monitored. Although the risk may be low, the potential consequences may be severe, and the impacts are difficult to assess in an Antarctic EIA.[184]

In these ways, the EIA requires extensive scrutiny by both the CEP and the ATCM itself, but the EIA does not effectively regulate tourism because it does not account for long-term environmental effects or assess the cumulative impact of human activities.

---

182 Daniela Liggett and Alan D. Hemmings, eds., "Exploring Antarctic Values: Proceedings of the Workshop Exploring Linkages Between Environmental Management and Value Systems: The Case of Antarctica," held at the University of Canterbury, Christchurch, New Zealand, 5 December 2011, available online at *http://antarctica-ssag.org/wp-content/uploads/2013/05/SSAG-proceedings-2013.pdf*, accessed 11 November 2013.

183 Daniela Liggett, "Tourism in the Antarctic: Modi Operandi and Regulatory Effectiveness," Ph.D. thesis, University of Canterbury, 2009, pp. 46–49.

184 Kees Bastmeijer and Ricardo Roura, "Regulating Antarctic Tourism and the Precautionary Principle," *American Journal of International Law* 98, no. 4 (October 2004): 763–781, quote from p. 770. For more on the implementation of regulations, see Mike G. Richardson, "Regulating Tourism in the Antarctic: Issues of Environment and Jurisdiction," in Davor Vids, ed., *Implementing the Environmental Protection Regime for the Antarctic* (2000).

Bastmeijer and Roura argue that there will be an uphill battle to ensure Antarctic preservation efforts into the future. This is largely because of the rise of eco-tourism and the radical expansion of the pleasure-cruise industry. Moreover, the Antarctic Treaty System has not implemented a full system for monitoring and auditing the results of human activity. There are also no worthwhile studies of the limits of tourism, their activities, and their impacts on the environment.[185]

What is clear is that there are very clear pressures on the system from tourism groups that will seek to exploit cracks in the Antarctic Treaty System for commercial gain. There is little question that tourism will eventually become a major economic activity in Antarctica. The core questions are what forms it will take—especially short visits versus four-star hotels and tourist compounds on the Antarctic coast—how it will be regulated, and how will it be logistically supported. A corollary question is, what technologies and methodologies will need to be pursued to ensure that this is done expeditiously and safely? This raises the specter of serious issues to be wrestled with in developing Antarctic tourism, some of them the same concerns that relate to eco-tourism in other places on Earth. Many of these issues have seemingly not been sufficiently understood and appreciated by those advocating for Antarctic tourism.

Central to this pressure on the Antarctic Treaty System, the National Science Foundation stands as the sole American entity overseeing U.S. efforts on the continent. It is the central organization that provides grants and contracts for activity in both the Arctic and Antarctic polar regions. NSF came into being in 1950 through a congressional act, largely at the behest of the scientific community. In particular, physicist Vannevar Bush published a study entitled *Science: The Endless Frontier*, which argued for an organization to coordinate basic science in America and science policy-making.[186] American involvement in the International Geophysical Year, a global civilian scientific initiative, inspired considerable interest in Antarctica. Thus, NSF created the United States Antarctic Program to coordinate American efforts on the continent. The structure of this program is as follows:

NSF Organization

National Science Foundation → Directorate for Geosciences (GEO) → Office of Polar Programs (PLR) → United States Antarctic Program (USAP)

Federal government interagency concerns affected the efforts in Antarctica in the late 1950s. IGY-Navy tensions emerged surrounding the command structure in Antarctica; for example, should a scientist lead, or perhaps a military officer should do so? NSF and the Navy settled on a dual-command system; despite

185 Bastmeijer and Roura, "Regulating Antarctic Tourism and the Precautionary Principle," pp. 763–781.

186 Vannevar Bush, *Science: The Endless Frontier, A Report to the President, Director of the Office of Scientific Research and Development, July 1945* (Washington, DC: Government Printing Office, 1945), available online at *https://www.nsf.gov/od/lpa/nsf50/vbush1945.htm,* accessed 24 July 2013. Also see Daniel Kleinman, *Politics on the Endless Frontier* (Durham, NC: Duke University Press, 1995). In a revisionist approach, Daniel Kevles decenters Bush as the impetus for NSF in "The National Science Foundation and the Debate over Postwar Research Policy, 1942–1945," *Isis* 68 (1977).

the rivalry, historian Dian Belanger argues that it proved a reasonable approach to governing activities on the continent.[187]

Private interests have driven one particular arena of Antarctic politics since the early Cold War: the exploitation of mineral and Earth resources. The Antarctic has been subject to exploitation since the 18th century, when explorers pursued sealing and whaling. By the 20th century, mineral resources had become a valuable pursuit, yet it is this arena that has largely been stifled from commercialization.

Throughout the 1980s, the several states with claims in the Antarctic debated the Convention on the Regulation of Antarctic Mineral Resource Activities. For roughly six years, 33 states debated the international agreement's provisions. Essentially, the bill would have legalized oil and mineral resource extraction upon the Antarctic continent. The United States and New Zealand, among others, expressed support for the bill. The State Department claimed "it is better to have regulated exploitation than a legal vacuum,"[188] in which no restraints had been formalized. However, the governments of Australia and France, among others, expressed deep environmental protectionism of the Antarctic continent and thus rejected the motion.

The Convention garnered 19 signatories, yet none ratified it, thus it never came into force. As such, there remained no tangible provision for environmental regulation in the Antarctic until the Protocol on Environmental Protection was added to the Antarctic Treaty in 1998; it received ratification by 33 states. Ultimately, the Protocol reserved "Antarctica as a natural reserve, devoted to peace and science."[189] In this regard, the United States can only promote certain types of private industry below the 60th parallel, given the international regulations it is subject to. Logistical support through contracts remains the primary opportunity for private ventures in the Antarctic.

In 1998, the U.S. government contracted with Air New Zealand to service aircraft for the U.S. Antarctic Program. The contract was essentially a five-year extension of the work Air New Zealand had already been conducting, but the U.S. Navy's decision to relinquish most logistical operations signaled a transition in the support program for American efforts on the continent. Instead, the Air National Guard and a contractor group, Antarctic Support Associates (ASA), provided the services previously handled by the Navy. NSF representative William Bryant announced that the contract presented the "most cost-effective and best operational option,"[190] which would provide for depot and vehicle maintenance and repair service for NSF facilities. Along with these benefits for the American program, Bryant stressed local benefits to the Christchurch, New Zealand, community, in the area of 20 to 26 million dollars. This contract signified not only the collaborative nature of American scientific initiatives abroad but also an opening for the private sector, given the withdrawal of the Navy. Antarctic Support Associates filled the void left by government displacement.

---

187 Dian O. Belanger, "The International Geophysical Year in Antarctica: A Triumph of 'Apolitical' Science, Politics, and Peace," in Launius, Fleming, and DeVorkin, eds., *Globalizing Polar Science*, p. 266. See also Dian O. Belanger, *Deep Freeze: The United States, the International Geophysical Year, and the Origins of Antarctica's Age of Science* (Boulder: University Press of Colorado, 2006).

188 Malcolm W. Browne, "France and Australia Kill Pact on Limited Antarctic Mining and Oil Drilling," *New York Times*, 27 September 1989.

189 "The Protocol on Environmental Protection to the Antarctic Treaty," 4 October 1991, available online at *http://www.ats.aq/e/ep.htm*, accessed 20 June 2013.

190 "NZ Wins $2m Contract for Ice Aircraft Work," *The Press* (Christchurch, NZ), 24 January 1998, p. 7.

Antarctic Support Associates is a Colorado-based civilian contractor. In October 1989, ASA submitted its proposal to NSF as a joint venture of Holmes & Narver Services, Incorporated (HNSI), and EG&G Incorporated (EGGI). HNSI is an Orange, California, management, architectural, and building services organization. EGGI is an Albuquerque, New Mexico–based technical services company specializing in systems engineering; integration; equipment design for command, control, and security systems; and data acquisition and monitoring. Both organizations have a history of government contracting, both locally and federally. The contract required ASA to support "operation of research laboratories, the procurement of equipment and supplies for laboratories and field parties of various sizes, design, procurement and construction of facilities."[191]

The agreement also specified that ASA would hold total operational accountability for both the Palmer and Amundsen-Scott South Pole stations as well as maintenance services for McMurdo Station. ASA also supplied travel for USAP and NSF personnel to and from Antarctica. The contract also left a provision for subcontracts from ASA. Moreover, ASA provided travel through the United States; Christchurch, New Zealand; and South America. NSF expected ASA to be in contact with the research projects and their investigators to ensure data and specimen delivery. NSF also expected ASA to provide general science support with vehicles, with data collection, and through meeting attendance. The contract also assigned ASA the task of infrastructure maintenance, such as plumbing, electrical service, power supply, scientific equipment, and housing. Essentially, NSF contracted out many of its operational objectives to ASA. The contract also specified the logistical support ASA was to provide.

Logistically, the contractors operated beyond the aforementioned Antarctic stations; the contract also supported the USAP from Port Hueneme near Oxnard, California, and in Punta Arenas, Chile, as well as Ushuaia, Argentina, and, as mentioned, in Christchurch, New Zealand. Each locale served a particular purpose of the operation. Port Hueneme functioned as the assembly and processing center, with office space and a warehouse. The port also served as a point of contact in the United States. Vendors in Punta Arenas, Chile, and Ushuaia, Argentina, supplied warehouse space, coordination, and support for vessel operations. Christchurch, New Zealand, operators also provided warehousing, cargo processing, and the facilitating of accommodations for American personnel.

In 2000, Raytheon Technical Services Company won a contract from the USAP to provide logistics, operations, and staffing to the NSF program, particularly at the McMurdo Station. Raytheon created a division specifically to carry out the 12-year contract called the Raytheon Polar Services Company (RPSC). The contract expired on 30 March 2012, fulfilling its duties to operate transport vessels, provide scientific support, and continually resupply the McMurdo Station.

Near the expiration of RPSC's contract in 2011, Lockheed Martin won a $2 billion Antarctic support contract from NSF. The contract began on 1 April 2012 and is set to last for a maximum of 4.5 years,

191 "Antarctic Support Associates Contract for Logistical Support to the U.S. Antarctic Program, NSF Contract No. DPP-8922632, available online at *http://www.nsf.gov/about/contracting/rfqs/support_ant/historical.jsp*, accessed 26 June 2013.

although it has milestones that, if not met, allow its cancellation. The competitive contract awarded by NSF mandates that Lockheed provide technological and infrastructural support to the USAP.[192]

Contemporary U.S. government policies, such as the various support service contracts awarded and the official end of the Navy's operation, suggest an increase in private involvement since the 1990s. In the 21st century, the National Science Foundation has been interested largely in finding contractors capable of providing transport services, facility maintenance, and operations and logistical support. Raytheon and then Lockheed took over many of the operations originally supported by the Navy. If the Coast Guard and Air National Guard were to similarly end their operations, it is plausible that NSF would contract their operations out to the private industry as well. The scientific investigations and research projects will remain the responsibility of American universities, research institutions, and federal agencies and will be provided for through government grants. The private industry, however, has increasingly assumed more logistical and infrastructural duties, absorbing them as contracts from the U.S. armed forces.[193]

Antarctica long has been a site of largely public interest rather than of private enterprise. I remain skeptical that the U.S. government is heavily interested in fostering commercialization. If the government is reluctant to encourage commercial enterprise in Antarctica, this could be due to a number of factors. There is considerable liability involved in an Antarctic undertaking, given the inhospitable conditions and geographic isolation. What I find most significant is the international community's privileging of environmental and scientific operations in Antarctica and in the preserving of the continent. Attempts to prevent the continent from being exploited by private natural-resource developers have used overwhelming international opposition to stifle a potentially profitable enterprise.

Also, environmentalist sentiment has prompted the regulation of the whaling and sealing industries. International regulations have subdued private involvement in resource exploitation overwhelmingly during the postwar years. While government contracting has fostered commercial interest in Antarctica since the 1980s, there is still an official shyness in fostering economic enterprises. Previously, the military supported the logistical and infrastructural operations of Antarctic exploration and human settlement without any private involvement whatsoever. This small movement toward increased competitive contracting for these operations to U.S. private industry is akin to the transition of air operations from a government entity to private-sector contractors. It portends the beginning of expanded economic activities in Antarctica.

As in that earlier air carrier transition in the 1920s, the U.S. government invested heavily in Antarctica to establish a lasting human presence. The military and the scientific community have also played a significant role, especially since the 1950s. Historian Marc Rothenburg states it best: "it would be incorrect, however, to assume that polar research truly became routine with better planning and the subsequent technological innovations of the 20th century."[194] Much of the success in corralling

---

192 Jeffrey Mervis, "Updated: NSF Picks Lockheed for Huge Antarctic Support Contract," ScienceInsider.com, 29 December 2011, available online at *http://news.sciencemag.org/2011/12/updated-nsf-picks-lockheed-huge-antarctic-support-contract*, accessed 11 November 2013.

193 "About That Antarctic Contract Rebid...," available online at *http://www.southpolestation.com/trivia/ncs/ncs.html*, accessed 11 November 2013.

194 Marc Rothenburg, "Making Science Global," in Launius, Fleming, and DeVorkin, eds., *Globalizing Polar Science*, p. 32.

Antarctica only for science research was because of the efforts of military leaders as facilitators, NSF as a funding source, and scientists as advocates. They did not, fortunately, have much in the way of organized opposition from commercial entities. Private industry was active in the region prior to World War II, but only in the niche markets of whaling and sealing. As international environmental regulations have restricted those practices as well as those of natural resource development, large sectors of private interests have been ruled out. Tourism and logistics, however, have become major avenues of Antarctic commercial activity.

Insisting that Antarctica remain a site of scientific inquiry with a limited commercial presence will ensure that support efforts will be limited to government contracts and the existing regulatory regime. It will also mean that scientific interest and overwhelming international environmentalist sentiment will be buttressed in efforts to defeat commercialization. Unless there is a significant change in international environmental policy toward the Antarctic continent, private interests may well be heavily restricted and/or regulated. That is not to say that the private sector has no place on the icy continent. Two significant areas of private involvement are active in Antarctica:

- Tourism: While the ATS is currently resolving how to regulate and monitor the effects of the tourist industry on the Antarctic environment and scientific pursuits, it remains an area of significant private interest represented by the International Association of Antarctica Tour Operators (IAATO), cruise lines, and travel agencies.
- Logistics: In the early 1990s, American contractors began to provide logistical, administrative, and technical support to governmental efforts.

The fact that this commercial activity has not progressed further is because of international, environmental, and industrial policy. The ATS's central tenets are threefold: safeguarding international peace, ensuring freedom of scientific research, and protecting the Antarctic environment. The third tenet of environmental protection became official with the 1991 adoption of the Protocol on Environmental Protection to the Antarctic Treaty. This provision officially disallowed the exploitation of Earth resources by mining interests. Other private industrial operations would require an environmental impact assessment (EIA), in particular for interference with "Antarctic flora and fauna, introducing non-native species into Antarctica, and entering Antarctic Specially Protected Areas."[195]

NASA's efforts in space closely mirror those of NSF in Antarctica. Both are inhospitable environments that are difficult to access. Both are zones protected from weapons, a Cold War trait, as well as mineral and other resource development. However, space, unlike Antarctica, is not strictly protected from environmental degradation. "Space junk" is becoming a concern for spaceflight, given the numerous decommissioned satellites, parts from multistage rockets, and many other pieces of floating debris. Anti-satellite tests, the Project Westford communications test, geostationary satellites, and other orbiting devices overpopulate both low-Earth orbit and geostationary orbit. In essence, there are numerous

195 Bastmeijer and Roura, "Regulating Antarctic Tourism and the Precautionary Principle," p. 768.

layers of space pollution. The U.S. military is monitoring the roughly 20,000 orbiting pieces of debris; however, there are no currently planned efforts to retrieve any of them.[196]

Thus, for the United States to promote space commercialization, it must make industry aware of the spatial conditions in which flight will take place. Spaceflight is already a risky business, but it is increasingly becoming more so with more and more devices flying over Earth in as many flightpaths. I recommend that the federal government begin to think about space debris and their corresponding flightpaths to ensure greater safety in space.[197] International organizations and national space organizations also should begin to think about regulating space as an environment prone to pollution and find a way to restrict which types of satellites should be orbited. Alternatively, the international community could develop a set of standards for satellites such as satellite durability and lifetime. This type of agreement will face harsh resistance but may well be necessary should orbital accidents begin to take place as we humans overpopulate our skies with satellites. If we avoid accidents not only during takeoff and landings but also during flight, space commercialization has a greater chance for success.

Antarctica has a legal status similar to that of the Moon. It is utilized primarily for scientific research, and no nation can claim its land. Yet basic supplies and logistic support for U.S. operations on the continent are provided by nonfederal organizations. Might this become a possibility in the future on the ISS or the Moon? Fostering such an approach to space activities could mean that control of orbital lunar assets would remain with NASA, which would select and fund science projects, oversee policy, and cycle personnel as necessary. The operation of these stations, however, could fall to a company with experience in remote locations, staffed by its own employees. Transportation to and from these stations could also be provided by outside organizations. At the same time, commercial activities could be encouraged.

Advancing Public Works

Frequently in the history of the United States, the federal government has developed critical infrastructure, often for its national security purposes, but then quickly transitioned it to the private sector for economic development. One of the most creative approaches to this process has been the use of the government or quasi-government development commission. There have been several of these in U.S. history. Among the most celebrated have been the Tennessee Valley Authority (TVA) and the Rural Electrification Administration, both of which originated as New Deal programs in the 1930s. Both of these efforts were viewed as a "public good" assisting in the enhancement of the general welfare of the

196 Nickolay Smirnov, *Space Debris: Hazard Evaluation and Mitigation* (Boca Raton, FL: CRC Press, 2002); Richard Talcott, "How We Junked Up Outer Space," *Astronomy* 36, no. 6 (June 2008): 40–43; Daniel Michaels, "A Cosmic Question: How To Get Rid of All That Orbiting Space Junk?" *Wall Street Journal*, 11 March 2009.

197 David Whitlock, "History of On-Orbit Satellite Fragmentations," NASA Johnson Space Center, 2004, available online at *http://www.orbitaldebris.jsc.nasa.gov/library/SatelliteFragHistory/13thEditionofBreakupBook.pdf*, accessed 11 November 2013.

people of the United States. Both served as investments that enabled regions that had not been developed to enhance economic competitiveness almost overnight.[198]

The Tennessee Valley Authority was created by an act of Congress on 13 May 1933 as part of President Franklin D. Roosevelt's New Deal agenda. The agency was to have the greatest effect on the daily life of Alabama residents. Many goals for TVA outlined by the act included flood control, improvement of navigation, reforestation of the lands in the valley, agricultural and industrial development, and the operation of defense plants at Muscle Shoals, a mass of jagged rocks that made boat passage impossible. By the late 1930s, TVA hydroelectric facilities had been built in Alabama. They were powered by the water behind three huge dams—Wilson, Wheeler, and Guntersville. Thereafter, electricity was available in places it had never been before and at rates that most local people could afford.[199]

As a federal corporation, TVA could not make a profit from its electric power sales. Its goal was to keep its electric power rates as low as possible while providing as much service as possible. Still a major entity in the Deep South, the agency's service area covers all of Tennessee and parts of Alabama, Mississippi, Georgia, Kentucky, Virginia, and North Carolina. Electricity from its plants flows through 17,000 miles of transmission lines to three million customers in an 80,000-square-mile service area. TVA's electrical power system remains the largest in the United States.

The establishment of TVA was only possible because of the Great Depression and the need to remediate economic suffering. As analyst Roscoe C. Martin concluded:

> At very long last, a combination of circumstances occurred which created a climate favorable to adoption of a public development measure. The principal elements which entered into the combination were the depression of the early nineteen-thirties, the depressed state of the Tennessee Valley, the continued support of Senator Norris, the leadership of President Roosevelt, and the state of mind of both the country and the organs of government, including principally and specifically the Congress. In the spring of 1933, President Roosevelt recommended to Congress action looking to the public development of the Muscle Shoals site. Senator George Norris and Representative (now Senator) Lister Hill joined in support of a measure which Congress accepted, and on May 18, 1933, the Tennessee Valley Authority Act became law.[200]

---

198 Basic histories of these organizations include North Callahan, *TVA: Bridge over Troubled Water, A History of the Tennessee Valley Authority* (New York: A. S. Barnes, 1980); Edwin C. Hargrove, *Prisoners of Myth: The Leadership of the Tennessee Valley Authority, 1933–1990* (Princeton, NJ: Princeton University Press, 1994); Richard A. Colignon, *Power Plays: Critical Events in the Institutionalization of the Tennessee Valley Authority* (Albany: State University of New York Press, 1996); Richard A. Pence and Patrick Dahl, *The Next Greatest Thing: Fifty Years of Rural Electrification in America* (New York: Consultants Library, 1984); Robert T. Beall, "Rural Electrification," *United States Yearbook of Agriculture* (Washington, DC: United States Department of Agriculture, 1940), pp. 790–809.

199 "Tennessee Valley Authority," *International Directory of Company Histories*, Encyclopedia.com, 2003, available online at *http://www.encyclopedia.com/doc/1G2-2845400109.html*, accessed 11 November 2013.

200 Roscoe C. Martin, "The Tennessee Valley Authority: A Study of Federal Control," *Law and Contemporary Problems* 22 (1957): 351–377, quote from p. 354.

Congress created TVA in an effort to utilize the natural resources of the valley to "promote the well-being of the people" and to "bring about healthy agricultural and economic conditions."[201]

Charles McCarthy, Assistant General Counsel of TVA in 1950, argued that since private companies would have no interest in the public purposes of flood control and navigation, it was necessary for TVA to have direct responsibility for developing the river. However, private industries and local governments had direct control over other resources in the valley, and TVA provided only technical assistance and encouragement. TVA could propose plans, but it could not enforce them, according to McCarthy, and had to rely on the "stimulation of local initiative." He added, "TVA has insisted…that the people of the Region must do the job. The TVA approach has been one of education, with reliance on demonstration rather than on compulsion."[202]

In terms of flood control, TVA had a powerful effect by harnessing the Tennessee River with nine dams and a management program aimed at making it an integral part of the Inland Waterways System. According to McCarthy, by 1950 the Norris Dam of 1936 had prevented no less than $41.5 million in flooding damage to Chattanooga, Tennessee. The system also fostered dramatic improvements to the agricultural community and created a well-received recreational system in the region. "More than $16,000,000 has been expended since 1933 for recreation facilities along TVA lake shores—for summer cottages, resorts, public parks, boats, mooring facilities, and the like," commented McCarthy. "The tourist industry, insignificant in 1933, is now recognized as the third industry of the Valley, amounting to some $175,000,000 annually." He also insisted that the contribution of TVA remade the Tennessee Valley into a desirable location for economic investment because of inexpensive electricity and the creation of a favorable business climate.[203]

TVA has been a public-private partnership from the beginning. The vast majority of its operations have been funded and managed by the government entity. The building of dams and other facilities, the production of electricity, and the delivery of other services have been largely the province of the government. Indeed, Roscoe Martin argued that it was an inherently governmental function:

> But the production of electricity is, by nature, a monopoly, whether a private or a public monopoly is not material: there is room for only one generating system in a given territory. The Act recognized this fact and directed the TVA to acquire whatever properties might be necessary to make its system effective. The TVA responded by negotiating the purchase of all generating systems within its "natural" power area. The last major purchase was completed in 1945, since which time the TVA has occupied a monopoly position with respect to the production of power in a region which covers some 92,000 square miles and which is, therefore, somewhat more than twice the size of the river basin."[204]

201 Charles McCarthy, "TVA and the Tennessee Valley," *The Town Planning Review* 21, no. 2 (July 1950): 116–130, quote from p. 117.

202 Ibid., pp. 117, 126.

203 Ibid., pp. 123, 129.

204 Martin, "The Tennessee Valley Authority," p. 363.

The private part of this partnership involved TVA's selling electricity, according to McCarthy, to "local distributors at wholesale rates under contracts which contain provisions requiring the distributor not to charge in excess of stipulated maximum retail rates and to devote to the reduction of rates surplus revenues which remain after meeting all obligations, including taxes in the case of co-operatives or tax-equivalent payments in the case of municipal distributors."[205] The private retailers were free to pursue their business model in the manner they deemed best. Through this relationship, TVA provided consumers with electricity at a rate lower than elsewhere in the nation and returned an average of 5 percent per year on the net investment in the TVA system.

The success of TVA prompted the creation of a similar effort elsewhere, overseen by the Rural Electrification Administration (REA). Taking electric power to the far reaches of the United States was unprofitable for private electricity firms, especially in far-flung rural areas of service. As late as the middle part of the 1930s, 9 out of 10 rural homes were without electrical service and had depressing living and working standards. In such settings, agriculture was the only viable economic activity, and even it was inefficient and less than fully profitable because of the lack of electrical power. Commenced on 11 May 1935, the REA began as a New Deal emergency measure and then expanded into an ongoing government activity, essentially offering electricity as a public utility. The REA undertook research to enhance the capacity and efficiency of electrical systems, pursued creative financing to allow private investors to build lines and offer services to farms, encouraged cooperatives to oversee the utility, and sought to turn the whole affair into a self-sustaining program. By the 1950s, more than 80 percent of all farms had electricity; by the 1960s, virtually all had power available. There have been efforts to make the system more efficient and less costly—with numerous reforms and regulations put into place to resolve these issues—but the overall result has been one of a strong public-private cooperative approach over the years.[206]

Frequently in the history of the United States, the federal government has developed critical infrastructure, often for its national security purposes, that quickly leads to economic development. At times, it has relied wholly upon private entrepreneurs. One of the most creative approaches to this process has been the use of the government or quasi-government development commissions to develop resources as a public good. There are many instances of this approach to public-private partnership. For example, in the General Mining Act of 1872, the U.S. government set up an uncontrolled but highly entrepreneurial structure that emphasized the principle that discovery conferred ownership. It left a legacy of riches and ruin that few wanted to repeat. More recently, commissions have been formed to create a more controlled development of the resource. Examples include the Isthmian Canal Commission (1904); the Bonneville Power Administration (1937); and the subject of this discussion, the Tennessee Valley Authority (1933). This entity was decentralized, not a conventional government agency. Congress provided at least the initial appropriation, but it was intended to become self-sustaining while delivering a public service. Corporate entities associated with it were empowered both to borrow and spend and to market goods and services.

205 McCarthy, "TVA and the Tennessee Valley," p. 122.

206 Bob Patton, "History of the Rural Electrification Industry," *Management Quarterly* 37, no. 4 (winter 1997): 2–7; Lowell J. Endahl, "Electrification of Rural America," in Gary Goreham, ed., *Encyclopedia of Rural America: The Land and People, Volume 1* (Santa Barbara, CA: ABC-Clio, 1997), pp. 223–227.

TVA served an important economic and social purpose and in the process served as the catalyst for the wholesale transformation of the region.

In the context of lunar development, might an organization similar to TVA be capable of commercially developing the Moon? Questions abound:

- Should it begin with the establishment of a lunar development commission/corporation?
- Would a commission/corporation start by building and managing lunar infrastructure for NASA?
- Would this be followed by an effort to spur economic development?

Making Accessible Scenic and Cultural Conservation Zones

A transformative and romantic perspective on nature transformed western civilization in the 19th century, and this has special application to the final case study on the preservation—but also accessibility—of American wilderness zones.[207] Efforts to preserve natural wonders are worth little unless people can visit and enjoy them. This prompted Americans to sponsor, through government action, a system of parks that would preserve for the future scenic wonder, make it available for broad visitation, and promote it as a means to individual health and recreation for the betterment of overall society. At the same time, the national and state governments and the lands they controlled took the prevention of private ownership of this land as a fundamental mission and as a means to ensure the land's preservation from ruthless exploitation. At the same time, the government sought to use private-industry services on government-held land so as to manage and maintain the integrity of the wilderness aesthetic while also fostering public access and some commercial opportunities. Initially, the American presidency (especially Presidents Ulysses Grant and Theodore Roosevelt) and Congress played key roles in the establishment of Yosemite and Yellowstone national parks. At the beginning of the 20th century, the federal government took action to establish the National Park Service (NPS) within the Department of the Interior, and it took responsibility both to preserve and to make accessible these parks.[208]

A significant case to be discussed is what is now known as Yosemite National Park, established as a state-owned and -operated entity by California in 1864, when Congress ceded the land to the state through a park bill that easily passed. The first Americans into the region realized that this was a place of exceptional beauty and sublime character. James Mason Hutchings, a "forty-niner" who entered the Yosemite Valley in 1851, realized that it held opportunity as a tourist destination. As one commentator put it:

> Hutchings immediately grasped the tourist potential in this landscape and set out to promote it as proof that America could compete with Europe for the traveler seeking scenic beauty. During the 1850's, the wealthy tourist who sought sublime scenery and mountain vistas headed for Switzerland, believing in the Romantic view that historical significance

207 One example of this transformation may be found in the manner in which Europeans responded to climate. See Vladimir Jankovi, *Confronting the Climate: British Airs and the Making of Environmental Medicine* (New York: Palgrave Macmillan, 2010).

208 This has been well analyzed in Roderick Nash, *Wilderness and the American Mind* (New Haven, CT: Yale University Press, rev. ed. 1982), especially pp. 120–141, 175–191, 205–207, 332–336.

> or evidence of a centuries-old civilization enhanced the beauty of a landscape. Yosemite lacked that kind of historical appeal, but the sharp rises of the mountains and the waterfall believed to be the highest in the world suggested that this landscape had more natural beauty to offer than the resort at the famed Swiss valley, the Lauterbrunnen. Hutchings began surveying the area for the construction of a hotel on the order of ones in Eastern resorts such as Niagara or the White Mountains in New Hampshire with the belief that if Americans were willing to pay for the falls at Niagara or the rolling mountains of New Hampshire, they would be eager to see American scenery on the order of Switzerland with a waterfall as high as nine Niagaras.[209]

Immediately, considerable private investment flowed into the valley. Several rail systems constructed spurs that brought tourists to visit the scenic wonder. This included not only the First Transcontinental Railroad (Sacramento to Stockton), but also a spur of the Central Pacific Railroad and the Yosemite Valley Railroad, both constructed to cash in on tourism to Yosemite.

Several companies supporting a growing tourism industry had established hotels, stores, and guide services by the turn of the 20th century. One of the first to do so was John Degnan, who, along with his wife, operated a bakery near the present-day Yosemite Lodge from the 1880s on. Near the same time, the Desmond Park Service Company, later renamed the Yosemite National Park Company (YNPC) in 1920, established numerous hotels, places to eat, and other services in the Mariposa Grove. The YNPC faced competition from the Curry Company, an outfit that offered rustic, low-cost lodging and dining at a camp called Camp Curry. The Curry Company eventually merged with YNPC to form the Yosemite Park & Curry Company. This came as part of the NPS's effort to limit concessionaires in national parks in the 1930s.[210]

In essence, these private-sector companies invested their own dollars to cash in on the rising tourism industry at Yosemite made possible by the maintenance of a pristine environment protected by the government. They served to make accessible to the public these scenic parks. The government *allowed* (and kept regulations to a minimum) these concessions, which started with a few proprietors who operated rustic lodges in the valley. As tourism grew and these facilities were strained, governmental managers had to step in to improve and increase the accommodations. They usually worked with the local proprietors to accomplish this task, sometimes providing seed money and low-interest loans to enhance the primitive hotels and campgrounds available. After World War II, they also had to manage rising expectations of visitors who sought more luxurious accommodations. Throughout this process,

---

209 "Discovery and Invention in the Yosemite," xroads, available online at *http://xroads.virginia.edu/~MA96/RAILROAD/yosemite.html*, accessed 24 November 2013.

210 Linda Wedel Green, *Yosemite: The Park and Its Resources; A History of the Discovery, Management, and Physical Development of Yosemite National Park, California* (Washington, DC: National Park Service, September 1987), pp. 133–152; Hans Huth, "Yosemite: The Story of an Idea," *Sierra Club Bulletin* 33 (March 1948): 63–76; Stanford E. Demars, *The Tourist in Yosemite, 1855–1985* (Salt Lake City: University of Utah Press, 1991); Joshua Johns, "All Aboard: the Role of the Railroads in Protecting, Promoting, and Selling Yellowstone and Yosemite National Parks," M.A. Thesis, University of Virginia, 1996; Alfred Runte, *Trains of Discovery: Western Railroads and the National Parks* (Los Angeles, CA: Roberts Rinehart, 1998).

NPS officials faced the challenge of preserving the scenic wonders while high tourist demand and rising expectations pressured the organizational structure that had been established early on. The balancing of Yosemite's protection with its development was more of an experiment than a well-thought-out strategy; nonetheless, the manner in which the housing and restaurant businesses have been incorporated into the park represents a set of divergent interests held in creative tension through a system of give and take as well as political pressure.[211]

Many individuals in government played significant roles in commercializing the parks, but it was the NPS that influenced them most. Between 1916 and 1933, NPS directors had a significant influence over the decisions in the parks. A succession of NPS directors held a corporatist attitude toward developing the park system so as to increase awareness and make the parks accessible to visitors.

Several scholars have characterized this collaborationist approach to public-private actions as the associative state. While bumps and bruises were inflicted on all sides during the 19th century as the industrial revolution reached its zenith, in the early 20th century a new associationalist approach to government involvement in business began to be played out. This approach legitimated the unification of industry and government to advance toward a too-often ill-defined but nonetheless positive goal, and the high priest of the philosophy was Secretary of Commerce Herbert Hoover in the 1920s. Because of the longtime relations between government and private industry, the parks offered a model for the associative state.[212]

Epitomizing this development, several National Park Service directors assisted in promoting the private activities at their preserves. Former businessman and Sierra Club member Steve Mather served as NPS director in the 1920s, establishing high standards among employees and administrators while promoting the parks. His successor, Horace Albright, continued Mather's efforts and began many of the NPS's historic preservation programs. The NPS leadership maintained its policies into the Depression and the ensuing New Deal era.[213]

Mather and Albright, as well as their successor, Arno Cammerer, promoted the parks through feature stories in publications such as *National Geographic*, the *Saturday Evening Post*, *Literary Digest*, and the *New York Times*. Historian Donald C. Swain argued that prior to 1933, the NPS courted political support in order to expand, arguing that "the purpose of the Park Service promotional campaigns was

211 Peter Blodgett, "Visiting 'The Realm of Wonder': Yosemite and the Business of Tourism, 1855–1916," *California History* 69, no. 2 (summer 1990): 118–133; Jonathan Spaulding, "Yosemite and Ansel Adams: Art, Commerce, and Western Tourism," *Pacific Historical Review* 65, no. 4 (November 1996): 615–639; Anne Hyde, "From Stagecoach to Packard Twin Six: Yosemite and the Changing Face of Tourism, 1880–1930," *California History* 69, no. 2 (summer 1990): 154–169; Alfred Runte, "Promoting Wonderland: Western Railroads and the Evolution of National Park Advertising," *Journal of the West* 31, no. 1 (January 1992): 43–48.

212 On Hoover and the "associative" state, see Louis Galambos, "The Emerging Organizational Synthesis in Modern American History," *Business History Review* 44 (autumn 1970): 279–290; Ellis W. Hawley, "Herbert Hoover, the Commerce Secretariat, and the Vision of an 'Associative State,' 1921–1928," *Journal of American History* 61 (June 1974): 116–140; Louis Galambos, "Technology, Political Economy, and Professionalization: Central Themes of the Organization Synthesis," *Business History Review* 57 (winter 1983): 471–493.

213 Donald C. Swain, "The National Park Service and the New Deal, 1933–1940," *Pacific Historical Review* 41, no. 3 (August 1972): 312–332.

largely economic. The objective was to benefit local economies near the parks, assist the concessioners in the parks…and indirectly stimulate the national economy."[214] In the 1930s, the NPS set up a Tourist Bureau based in New York City to foster interest via travel agencies and interest groups. The Tourist Bureau symbolized the efforts of the government to encourage interest in the parks; however, it was a short-lived operation that dissolved when Congress authorized the U.S. Travel Bureau in 1940.[215]

Also, through the 1930s, the Civilian Conservation Corps presented a unique opportunity for the Park Service. It allowed the agency to hire young men to work in the parks, open new parks, and promote the existing parks. However, the heavy promotional efforts fomented a negative response from some park-goers. Secretary of the Interior Harold Ickes, feeling pressure from wilderness preservationists, began to move away from accessibility in favor of preservation. He disliked the overly crowded national parks and sought to build fewer roads in them due to the disturbance of automobiles. This sentiment influenced the selection of Cammerer's successor in 1940, Newton Drury, a strong conservationist. Drury's directorship ended the "Mather tradition" of greater access and services and emphasized greater synchronization with the efforts of the Sierra Club, the Wilderness Society, and the National Parks Association.[216]

In the latter third of the 20th century, however, pressures on the system forced first a return to the earlier associative state of public-private partnership and then ever-greater reliance on private investment to operate amenities in the national parks. John L. Crompton, a social science researcher in parks and recreation, suggests four related macro issues that have sparked this transition. First, reductions in the public funding levels traditionally available for the National Park Service have compelled greater reliance on private investment and a corresponding openness to business interests in the parks. The erosion of the public investment comes from several sources, according to Crompton: "Pragmatists seek a more effective government and see privatization as a means to that end. Commercial interests seek to obtain more business by taking over some of the agency's financing, production, or operating roles. For ideologues, privatization is a political agenda aimed at ensuring that government plays a smaller role compared to private institutions."[217]

Second, left-of-center populists also often view privatization as a positive, for it gives external forces "greater power to satisfy their common needs, while diminishing that of larger public bureaucracies." Third, Crompton emphasizes the place of inefficiencies present in any monopoly, either public or private, in motivating a transition to more private delivery of services in the national parks. "In many communities, however, a park or recreation agency is a monopoly supplier of many services. Privatization is seen as a means of inducing competition into public agency monopoly situations."[218] Finally, often park managers recognize that they do not have the expertise in-house to provide many key services needed

214 Ibid.

215 Ibid.

216 Ibid.

217 John L. Crompton, "Forces Underlying the Emergence of Privatization in Parks and Recreation," *Journal of Park and Recreation Administration* 16, no. 2 (1998): 88–101, quote from p. 88.

218 Ibid., p. 88.

by users of the park. Rangers, for example, are not restaurateurs, and it makes sense to contract out that function at virtually every park.

The result of this development has been that the National Park Service has long enjoyed a public-private partnership in the delivery of services to visitors going back to the 19th century. This cooperative approach reached a high-water mark in the Progressive Era of the early 20th century through the associative state model of action. During the New Deal, as the federal government took control of more aspects of the nation's resources, this relationship languished, but it returned to the norm in the latter part of the 20th century. Into the 21st century, however, the balancing of public desire to preserve scenic wonders versus enhancing commercial activities remains a task not without difficulties.

An example of this may be found in the white-water rafting experience of the Grand Canyon. In 1972, 16,432 people rafted the Colorado River through the Grand Canyon, with a variety of companies providing this service to all comers. In the process, the human impact on the site proved disastrous to the wilderness setting, an obvious conclusion when more than 500 rafters were leaving the Lee's Ferry launch site for a trip down the river every day of the season. The Park Service sought to regulate this situation by freezing visitation at 1972 levels and prohibiting large numbers of motor vessels, controlling the amount and types of equipment and supplies, and apportioning daily launches. This has proven a controversial regulatory experience, and there have been numerous forays and rebuttals by various groups to manage this unique resource, with no end to the controversy in sight.[219]

As historian Roderick Nash has concluded for this case (and it is applicable elsewhere as well): "The root of the problem, of course, was that the public had become too large for the wilderness."[220] In addition, the public had different expectations, and companies sought to provide experiences commensurate with those expectations. For those who wanted to see the sights but were more interested in beer and beach volleyball than in an encounter with the wilderness, motorboats on the Colorado River were optimal. For purists who wanted to experience the wilderness setting untouched by human incursion, those other users were anathema. Motor vessels and drunken parties destroyed the wilderness experience for those seeking that experience. In both cases, companies were filling a need, so it was not simply a fight between preservationists and developers, although that was also a part of the debates on many occasions. Perhaps the fundamental nature of democratic institutions is that controversies proliferate and are never fully resolved, and if so, this sort of give and take is a good lesson in the structure of democratic society.

In terms of applicability to the space frontier, the experience of the National Park Service is most germane in terms of space tourism efforts. When Congress created the U.S. National Park Service in 1916 to conserve natural and historical resources "by such means as will leave them unimpaired," a key component was to assist the public in reaching those scenic wonders. Accordingly, park managers, recognizing the need for public support to encourage future preservation, allowed private entrepreneurs to commercialize

219 W. E. Garrett, "Grand Canyon: Are We Loving It to Death?" *National Geographic* 154 (1978): 16–51; Robert Dolan, Alan Howard, and Arthur Gallenson, "Man's Impact on the Colorado River in the Grand Canyon," *American Scientist* 62 (1974): 392–401; "Troubled Waters," *Newsweek* 81 (1973): 62.

220 Nash, *Wilderness and the American Mind*, p. 337.

the parks in such a manner as to encourage public visitation. They encouraged railroad companies and other concessionaires to build hotels and related facilities in the national parks. Those concessionaires then paid fees that the National Park Service used to build additional roads and trails. Americans rode and drove to the national parks, vastly expanding tourism and creating the family vacation tradition.

Government officials, as well as policy, could likewise encourage private-sector development in space tourism, both in low-Earth orbit and on the Moon. The following possibilities exist:

- Public officials could expand the use of government facilities by private entrepreneurs as a means of encouraging public use and visitation.
- Private firms could pay fees that government agencies could then use to expand and develop facilities.
- Government could create a favorable regulatory climate for space tourism.
- Private citizens could then experience space through both remote access and direct participation.

Beyond these very specific possibilities, NASA could also award lease contracts for habitation/support services of facilities in orbit and on the Moon. Baseline development and operational costs could then be funded by NASA lease. As an additional revenue stream, companies could then add tourism for marginal costs. Such an environment could create a public-private space ecology with efficiencies of operations achieved through economies of scale.

Conclusion

For human spaceflight to expand into the future, there must be a compelling reason to undertake it. From the defining event of Sputnik in 1957, I would contend that there have been five major themes—and only these five—that have been effective in justifying a large-scale, publicly funded spaceflight agenda:

1. scientific discovery and understanding
2. national security and military applications
3. economic competitiveness and commercial applications
4. human destiny/survival of the species
5. national prestige/geopolitics

Specific aspects of these five rationales have fluctuated over time but remain the only reasons for the endeavor that have any saliency whatsoever.[221]

The first and most common rationale for spaceflight is that an integral part of human nature is a desire for discovery and understanding. At one level, there exists the ideal of the pursuit of abstract scientific knowledge—learning more about the universe to expand the human mind—and pure science and exploration of the unknown will remain an important aspect of spaceflight well into the foreseeable future. This goal clearly motivates the scientific probes sent to all of the planets of the solar system. It propels a wide range of efforts to explore Mars, Jupiter, and Saturn projected for the 21st century. It energized such efforts as the Hubble Space Telescope, which has revolutionized our knowledge of the universe since its deployment in 1990.[222]

A second rationale of national defense and military space activity has also proved useful for spaceflight advocates. From the beginning, national leaders sought to use space to ensure U.S. security against nuclear holocaust. For instance, in 1952, a popular conception of the U.S.-occupied space station showed it as a platform from which to observe the Soviet Union and the rest of the globe in the interest of national security. The human spaceflight enterprise also gained energy from Cold War rivalries in the 1950s and 1960s as international prestige, translated into American support from nonaligned nations, found an important place in the space policy agenda. Human spaceflight also had a strong military nature during the 1980s, when astronauts from the military services deployed reconnaissance satellites into Earth orbit from the Space Shuttle.[223]

---

221 I have wrestled with this question in Roger D. Launius, "Compelling Rationales for Spaceflight? History and the Search for Relevance," in Steven J. Dick and Roger D. Launius, eds., *Critical Issues in the History of Spaceflight* (Washington, DC: NASA SP-2006-4702, 2006), pp. 37–70.

222 An excellent discussion of all space probes launched to date may be found in Asif A. Siddiqi, *Deep Space Chronicle: Robotic Exploration Missions to the Planets* (Washington, DC: NASA SP-2002-4524, 2002). See also Robert W. Smith, *The Space Telescope: A Study of NASA, Science, Technology, and Politics* (New York: Cambridge University Press, 1989); David H. DeVorkin and Robert W. Smith, *The Hubble Space Telescope: Imaging the Universe* (Washington, DC: National Geographic, 2004).

223 See Paul B. Stares, *The Militarization of Space: U.S. Policy, 1945–1984* (Ithaca, NY: Cornell University Press, 1985); Damon Coletta and Frances T. Pilch, eds., *Space and Defense Policy* (London, U.K.: Routledge, 2010); James Clay Moltz, *The Politics of Space Security: Strategic Restraint and the Pursuit of National Interests* (Stanford, CA: Stanford University Press, 2011, second edition).

The third rationale of economic competitiveness and commercial applications also represents a useful role that the public accepts for spaceflight. Space technologies, especially the complex human spaceflight component, demand a skilled and well-trained workforce whose talents are disseminated to the larger technological and economic base of the nation. The Apollo program, for example, served explicitly as an economic engine fueling the southern states' economic growth. In recent years, the economic rationale has become stronger and even more explicit as space applications, especially communications satellites, become increasingly central for maintaining United States global economic competitiveness.[224]

The fourth imperative for spaceflight has revolved around human destiny. With Earth so well known, advocates argue, exploration and settlement of the Moon and Mars is the next logical step in human exploration. Humans must question and explore and discover or die, advocates for this position insist. There is also a terrifying aspect to this rationale: humanity will not survive if it does not become multiplanetary.[225]

Finally, national prestige and concern for geopolitical relations has dominated so many spaceflight decisions that it sometimes seems trite to suggest that it has been an impressive rationale over the years. Yet there is more to it than that, for while all recognize that prestige sparked and sustained the space race of the 1960s, we too often fail to recognize that it continues to motivate support for NASA's programs. The United States went to the Moon for prestige purposes, but it also built the Space Shuttle and embarked on the Space Station for prestige purposes.[226]

Of these five rationales, the latter two have receded into the background as drivers for sustained public investment in the American space program. National security and science are important drivers—but not for the human space agenda because they are oriented toward robotic efforts. Commercial activities are the primary arena that might prove successful for energizing human spaceflight.

There is really no other way, it seems to me, as this sermon on the mount makes clear. The national debt of the United States in 2013 stood at $17.07 trillion. Americans add to this every year, and the national debt is rising at an unsustainable rate. This is in no small measure because the two largest expenditures for the United States on an annual basis—national defense and entitlements (including Social Security, Medicare and Medicaid, and other payments to specific constituencies)—account for more than half of all annual expenditures in the federal budget. The income received by the U.S. government in fiscal year 2013 was

224 Roger D. Launius, "Lessons from Terrestrial Exploration for Earth Orbit," *Space News* 24, no. 41 (21 October 2013): 19, 21. See also Roger Handberg, *International Space Commerce: Building from Scratch* (Gainesville: University Press of Florida, 2006); Lawrence A. Cooper, "Encouraging Space Exploration Through a New Application of Space Property Rights," *Space Policy* 19 (May 2003): 111–118; Alan Wasser and Douglas Jobes, "Space Settlements, Property Rights, and International Law: Could a Lunar Settlement Claim the Lunar Real Estate It Needs To Survive?" *Journal of Air Law and Commerce* 73 (winter 2008): 37–78.

225 Carl Sagan, *Cosmos* (New York: Random House, 1980), pp. 231–232; John W. Young, "The Big Picture: Ways To Mitigate or Prevent Very Bad Planet Earth Events," *Space Times: Magazine of the American Astronautical Society* 42 (November–December 2003): 22–23. See also "Human Exploration and Development of Space (HEDS) Enterprise," 30 October 1997, available online at *http://www.hq.nasa.gov/office/nsp/heds.htm*, accessed 8 March 2012.

226 See Roger D. Launius, "Imprisoned in a Tesseract: NASA's Human Spaceflight Effort and the Prestige Trap," *Astropolitics: The International Journal of Space Politics & Policy* 10, no. 2 (May–August 2012): 152–175.

$2.77 trillion, approximately $230.8 billion each month. Expenditures in fiscal year 2013, however, were approximately $3.45 trillion per year, some $287.5 billion each month. During that fiscal year, spending exceeded income by $680.3 billion; this was good news because it was the smallest gap in income/expenditures since fiscal year 2008, and it almost halved the $1.09 trillion shortfall seen in fiscal year 2012.[227]

While those FY 2013 figures represented an important step forward, the longstanding shortfall of income versus expenditures in the United States has been a serious issue for some time and may approach crisis proportions in the next decade or so. Something has to give, and hard choices will have to be made. The space program has long consistently ranked near the top of those activities of the federal government deserving to be cut. In the 1960s, during the height of the Apollo program, most Americans preferred spending those same dollars for a range of activities—mitigation of air and water pollution, job training for unskilled workers, a war on poverty, etc.—rather than NASA.[228]

Ironically, at the time of the excitement of the Apollo 11 Moon landing, a Harris Poll in 1969 reported that 56 percent of Americans believed that the costs of the Apollo program were too great and that 64 percent believed that $4 billion a year for NASA was too much.[229] The National Science Foundation contracted with the Opinion Research Corporation for public opinion studies in 1972, 1974, and 1976, and each found that the space program was at the bottom of 13 items desired by Americans in terms of science and technology investment. The same is true of a 2007 Harris poll in which this question was asked: "If spending had to be cut on federal programs, which two federal program(s) do you think the cuts should come from?" Tops on the list was the space program, which garnered 51 percent advocating its reduction, followed by welfare (28 percent), defense spending (28 percent), and farm subsidies (24 percent).[230]

In such an environment, any expansive space program in the 21st century will require a practical, cost-effective, commercial basis. Leveraging a declining public investment for this agenda with public-private partnerships on a more equitable basis is the most obvious methodology for achieving an expansive future in space. While one can question an emerging neoliberal perspective that argues that government support of scientific research is counterproductive to wealth-generating technology and that private enterprise can supply most if not all of the funds required for both pure and applied research, there is still no doubt that less funding will be available for this endeavor in the future than in the past.[231]

227 "Budget Deficit in the U.S. Narrows to 5-Year Low on Record Revenue," *Bloomberg News*, 30 October 2013, available online at *http://www.bloomberg.com/news/2013-10-30/budget-deficit-in-u-s-narrows-to-5-year-low-on-record-revenue.html*, accessed 14 December 2013; Alex Zhavornkov, *The Ageless Generation: How Advances in Biomedicine Will Transform the Global Economy* (New York: Palgrave Macmillan, 2013), pp. 145–147.

228 Stephanie A. Roy, Elaine C. Gresham, and Carissa Bryce Christensen, "The Complex Fabric of Public Opinion on Space," *Acta Astronautica* 47 (2000): 665–675.

229 Louis Harris, *The Harris Survey Yearbook of Public Opinion, 1970* (New York: Louis Harris and Assoc., 1971), pp. 83–84. See also Timothy B. Kyger, "President Nixon, Public Opinion, and Post-Apollo Planning," Senior Thesis, University of San Francisco, 1985, pp. 53–56.

230 "Closing the Budget Deficit: U.S. Adults Strongly Resist Raising Any Taxes Except 'Sin Taxes' or Cutting Major Programs," *Harris Poll #30*, 10 April 2007.

231 See Terence Kealey, *The Economic Laws of Scientific Research* (New York: Palgrave Macmillan/St. Martin's Press, 1996).

There are many possibilities for pursuing public-private partnerships for space exploration. The National Council for Public/Private Partnerships offered a list of seven best practices for any effort to pursue these relationships:

1. **PUBLIC SECTOR CHAMPION:** Recognized public figures should serve as the spokespersons and advocates for the project and the use of a PPP. Well-informed champions can play a critical role in minimizing misperceptions about the value to the public of an effectively developed PPP.

2. **STATUTORY ENVIRONMENT:** There should be a statutory foundation for the implementation of each partnership. Transparency and a competitive proposal process should be delineated in this statute. However, unsolicited proposals can be a positive catalyst for initiating creative, innovative approaches to addressing specific public sector needs.

3. **PUBLIC SECTOR'S ORGANIZED STRUCTURE:** The public sector should have a dedicated team for PPP projects or programs. This unit should be involved from conceptualization to negotiation, through final monitoring of the execution of the partnership. This unit should develop Requests For Proposals (RFPs) that include performance goals, not design specifications. Consideration of proposals should be based on best value, not lowest prices. Thorough, inclusive Value for Money (VfM) calculations provide a powerful tool for evaluating overall economic value.

4. **DETAILED CONTRACT (BUSINESS PLAN):** A PPP is a contractual relationship between the public and private sectors for the execution of a project or service. This contract should include a detailed description of the responsibilities, risks and benefits of both the public and private partners. Such an agreement will increase the probability of success of the partnership. Realizing that all contingencies cannot be foreseen, a good contract will include a clearly defined method of dispute resolution.

5. **CLEARLY DEFINED REVENUE STREAM:** While the private partner may provide a portion or all of the funding for capital improvements, there must be an identifiable revenue stream sufficient to retire this investment and provide an acceptable rate of return over the term of the partnership. The income stream can be generated by a variety and combination of sources (fees, tolls, availability payments, shadow tolls, tax increment financing, commercial use of underutilized assets or a wide range of additional options), but must be reasonably assured for the length of the partnership's investment period.

6. **STAKEHOLDER SUPPORT:** More people will be affected by a partnership than just the public officials and the private sector partner. Affected employees, the portions of the public receiving the service, the press, appropriate labor unions and relevant interest groups will all have opinions, and may have misconceptions about a partnership and its value to all the public. It is important to communicate openly and candidly with these stakeholders to minimize potential resistance to establishing a partnership.

7. **PICK YOUR PARTNER CAREFULLY:** The "best value" (not always lowest price) in a partnership is critical in maintaining the long-term relationship that is central to a successful partnership. A candidate's experience in the specific area of partnerships being considered is an important factor in identifying the right partner. Equally, the financial capacity of the private partner should be considered in the final selection process.[232]

With this as the case, are there lessons from the past six case studies that might be applied to future public-private partnerships in space? In the context of these historic examples of public-private partnerships, the most applicable to space activities are in transportation, and I will concentrate me comments on this arena: railroads and aviation and their relationship to space transportation. Public-private partnerships in the development of American transportation systems have been varied, often complex, and, over time, remarkably boutique. Transportation partnerships have gradually evolved, taken a divergent set of paths, and ranged from fully public to fully private and virtually everything in between. Only one conclusion may be reached concerning the development of these railroads—whether the initial transportation mode began as a private or a public initiative—all successful railroads have incorporated a mixed model of funding and operations.

Initial railway development began as a private enterprise, but the costs of investment were too great to be sustained. This led to the entrance of the government—sometimes local, often state, and, in the latter 19th century, federal—to underwrite the cost of investment in a variety of inventive ways ranging from direct ownership to land grants, regulatory reforms, tariff splits, bond sales, and the like. Regardless of the public investment, private enterprise tended to dominate the public-private partnership. By the latter 19th century, rail systems had grown so critical to American economic expansion, national security, and migration that the federal government intervened to assert greater regulatory power over this partnership, regulating services and costs, standardizing systems, and enforcing safety. Most important, it never asserted ownership—in contrast to other models in other nations—except in times of war.

The development of the airline industry followed a different path, but one also related to the lesson learned in the railroad era. The industry initially began as a largely governmental function, but over the course of the 20th century, it was privatized more and more, with the government serving as regulator and facilitator—regulator in terms of safety and process, and facilitator in terms of advocating for the cause of flight. Indeed, the federal government, more than any other entity, made possible the formation of the public-private partnerships that eventually have come to dominate the air transport industry. At the same time, in the latter third of the 20th century, the industry became increasingly free of regulation, allowing it to pursue business decisions enhancing its income.

In terms of the space transportation industry, there is a similar opportunity to learn from the air transportation industry. The dominance of the U.S. government, in the form of NASA, is receding into the background as the 21st century progresses. A major step in that process was the 2011 retirement of the Space Shuttle, a $3 billion–a–year program whose cancellation forced NASA to seek other solutions to

232 "7 Keys to Success," National Council for Public-Private Partnerships, available online at *http://www.ncppp.org/ppp-basics/7-keys/*, accessed 24 November 2013.

space transportation. This is an overwhelmingly positive development, for funding from that program has been invested in kick-starting commercial space access efforts. In a public-private partnership to develop new low-Earth orbit capabilities, four major firms are using government and their own entrepreneurial dollars to solve this problem. The success of at least some of these firms seems assured. The government role, beyond some investment, is primarily focused on oversight and regulation. A major challenge will be to negotiate the relationship between safety and aggressive operations.

The government has succeeded thus far in sponsoring a moderate path forward for space commercialization. No other nation has been able to chart this course. The passing and signing of The Commercial Space Launch Amendments Act of 2004, H.R. 3752, for example, enabled the FAA to license and regulate commercial human spaceflight while providing liability protection for companies. This legislation will be critical to the success of commercial space transport. In addition, NASA's financial rewards, prizes, and other incentives, as well as technical assistance to emerging private companies, are just as central to the way forward.

I was struck by the film *2001: A Space Odyssey*, as spacefarers shuttled from Earth to an orbiting space station aboard a commercial space plane and private firms carried out many other functions in low-Earth orbit, including the Hilton hotel on the station. That always seemed to me the appropriate structure for operations in orbit.[233]

I was equally struck by the truism that the Spanish monarchs Ferdinand and Isabella funded Christopher Columbus's voyages of discovery as a national program for the good of their emerging European power. It is also impossible to get past the point that most other voyages of discovery were sponsored largely by a sovereign. In every case, I hasten to note, these European monarchs sought tangible results—usually wealth, often geopolitical advantage, and sometimes less-specific positive outcomes—that accrued to the nation through the exploration paradigm.

The two models were very different, one a government program and the other a private-sector effort. Both models are closer to each other than might be expected because of the public-private partnerships that made them possible. What has happened, however, is that certain parts of space are now well incorporated in the normal realm of human operations, just as European explorers incorporated the Western Hemisphere into their normal realm of activities. In an irony of the first magnitude for those who insist that humans are not truly engaged in space exploration because they have been confined to low-Earth orbit since the end of Apollo more than 40 years ago, the Space Shuttle turned orbital space into a place that was no longer a frontier. The first astronauts and cosmonauts, of course, truly were explorers in the traditional sense of the term because they pushed back the frontiers of knowledge about this unusually strange, harsh, and hostile environment. They learned how to operate there and to do useful things in the region.

I will give those with the "right stuff" of the 1960s high marks for pioneering operations in space. But I will give the astronauts and cosmonauts of a later era—those flying on the Space Shuttle, Mir, and the

233 This section is based on Roger D. Launius, "Lessons from Terrestrial Exploration for Earth Orbit," *Space News* 24, no. 41 (21 October 2013): 19, 21.

International Space Station—equally high marks for turning orbital operations into routine, useful activities. There is no longer any mystery about what we will experience in this region, and we understand well how to use it effectively and economically for a range of activities. This was what happened in terrestrial exploration as well: government-funded pathfinders preceded those engaged in a range of commercial and other private activities in the new region.

That brings me back to *2001: A Space Odyssey*. What has been the norm in human history is that as explorers have turned unknown regions into known ones, exploration has given way to frontiering. We are in the frontiering stage of low-Earth orbit, and there are all manner of possibilities for exploiting it for commercial purposes. The range of possibilities is breathtaking: commercial microgravity research on all manner of products, applications both well known and still in the process of becoming, and even tourism. The efforts of Virgin Galactic to create a suborbital space tourism capability is well known and will almost certainly find realization in the near term, but the partnering of Bigelow Aerospace and SpaceX for orbital tourism also has potential. Who knows what might flow from commercial beachheads in Earth orbit?

Most assuredly, we will see more entrepreneurial activity in Earth orbit in the next few years. New and unexploited regions have always attracted entrepreneurs—some succeed and others do not—and orbital space does not look particularly different from previous terrestrial frontiers in that regard.

Flash forward to 2050 and imagine astronauts boarding a commercial space launcher delivering crews to a space station or a lunar base. This may become reality for space exploration if commercial activities in space advance. Of course, this new direction proposes a major shift in the way NASA accomplishes human spaceflight. In this context, NASA could return to its roots as a research-and-development organization pursuing transformational technologies while private firms operate space systems. Turning low-Earth orbit over to commercial entities empowers NASA to focus on deep space exploration, perhaps eventually visiting Mars.[234]

To make this a reality, NASA must undergo significant shifts in programs and practices. For many years, NASA ran things as it saw fit. While it relied on contractors for assistance, no one doubted who was in charge. This has to change; NASA must now develop more equal partnerships to accomplish its space exploration mandate. This shift will be difficult, but it is critical for a robust future. The time has arrived for NASA to shift from building and operating space launch systems to purchasing these services from commercial firms.

Further, utilizing America's emerging space transportation companies will ignite ingenuity and innovation. It will result in multiple commercial firms offering space launch services, offering redundancy in space access. Competition will help reduce the cost while enhancing capability. This new approach could extend the lifespan of the International Space Station as the greatest engineering asset the world has ever jointly developed. Now poised to accomplish much during the next 10 years in the arenas of

234 This section is based on Roger D. Launius and G. Michael Green, "New Vision for NASA," *Florida Today*, 14 February 2010.

biomedical science, Earth science, and materials research, the ISS also offers a unique capability to learn how humans function during long-duration space missions.

At the same time, this new path need not abandon human exploration; it could provide the United States with a sustainable and executable space exploration plan indefinitely. It furthers human activities on the ISS and opens translunar space sooner than other approaches—on American rockets built by commercial providers.

Selective Annotated Bibliography

Key Historical Studies

Ambrose, Stephen E. *Nothing Like It in the World: The Men Who Built the Transcontinental Railroad 1863–1869.* New York: Simon & Schuster, 2000. This is the classic story of the building of the railroads.

Bain, David Haward. *Empire Express: Building the First Transcontinental Railroad.* New York: Viking Penguin, 1999. This is another good account of the building of the transcontinental railroad.

Beebe, Lucius. *The Central Pacific & the Southern Pacific Railroads: Centennial Edition.* Los Angeles, CA: Howell-North, 1969. A specialty work.

Belanger, Dian Olson. *Deep Freeze: The United States, the International Geophysical Year, and the Origins of Antarctica's Age of Science.* Boulder: University of Colorado Press, 2006. Belanger provides a very fine institutional history of the National Science Foundation's efforts to create and sustain the American foothold in Antarctica.

Brooks, John. *Telephone: The First Hundred Years.* New York: Harper & Row, 1976. A standard work.

Bruce, Robert V. *Bell: Alexander Bell and the Conquest of Solitude.* Ithaca, New York: Cornell University Press, 1990. This is possibly the best biography of the inventor of the telephone.

Cooper, Bruce C. *Riding the Transcontinental Rails: Overland Travel on the Pacific Railroad 1865–1881.* Philadelphia: Polyglot Press, 2005. This work is a study of how the railroad opened the American West.

Cooper, Bruce Clement, editor. *The Classic Western American Railroad Routes.* New York: Chartwell Books/Worth Press, 2010. Cooper provides a discussion of the routes of the trains.

Creese, Walter L. *TVA's Public Planning: The Vision, the Reality.* Knoxville: University of Tennessee Press, 1990. This work stresses the utopian goals of the program.

Goodrich, Carter. *Government Promotion of American Canals and Railroads, 1800–1890.* Westport, CT: Greenwood Press, 1960. The author has written an indispensable account of government's role in fostering transportation.

Gudmundsen, Sveinn Vidar. *Flying Too Close to the Sun: The Success and Failure of the New-Entrant Airlines.* Burlington, VT: Ashgate Publishing, 1999. This work is an important study of recent markets in air transportation.

Hubbard, Preston J. *Origins of the TVA: The Muscle Shoals Controversy, 1920–1932.* Nashville, TN: Vanderbilt University Press, 1961. This work is a critical assessment of the manner in which the program was inaugurated.

Larson, John Lauritz. *Internal Improvement: National Public Works and the Promise of Popular Government in the Early United States*. Chapel Hill: University of North Carolina Press, 2001. Larson's work is a good study of early efforts to foster economic development.

—. *The Market Revolution in America: Liberty, Ambition, and the Eclipse of the Common Good*. New York: Cambridge University Press, 2009. Here, the author captures the American ambivalence about the emergence of modern capitalism.

Launius, Roger D., James Rodger Fleming, and David H. DeVorkin, editors. *Globalizing Polar Science: The International Polar Years and the International Geophysical Year*. New York: Palgrave Macmillan, 2010. This is a collection of key articles on the importance of international scientific efforts at the poles.

Lively, Robert A. "The American System: A Review Article." *Business History Review* 29 (March 1955): 81–96. This work is a recommended starting point.

Neuse, Steven M. "TVA at Age Fifty: Reflections and Retrospect." *Public Administration Review* 43 (Nov.–Dec. 1983): 491–499. Neuse's work is a useful short overview.

Pyne, Stephen E. *The Ice: A Journey to Antarctica*. New York: Ballantine Books, 1988. Pyne touches on the elements of Antarctica and emphasizes it as a realm of history, a place of dominant landscapes, an inspiration for literature and art, and a scientific treasure trove. He argues that for all of its geology and geography, geomagnetism and weather, biology and boredom, Antarctica remains at sum a diminished location in which water has been transformed into mineral. There are no cultural studies beyond those on the explorers themselves.

Runte, Alfred. *National Parks: The American Experience*. Lincoln: University of Nebraska Press, 1997, 3rd ed. This useful short history provides an overview the National Park Service and its parks.

Shapiro, Edward. "The Southern Agrarians and the Tennessee Valley Authority." *American Quarterly* 22 (winter 1970): 791–806. This work outlines how southern conservatives supported TVA as a counterpoint to northern big business's incursion into the region.

Van der Linden, F. Robert. *Airlines and Air Mail: The Post Office and the Birth of the Commercial Aviation Industry*. Lexington: University Press of Kentucky, 2002. This is an important study of government fostering of the airlines.

Wirth, Conrad L. *Parks, Politics, and the People*. Norman: University of Oklahoma Press, 1980. Wirth provides a scholarly history of the nation's parks.

Wu, Tim, *The Master Switch: The Rise and Fall of Information Empires*. New York: Alfred A. Knopf, 2010. Wu has written an excellent overview.

Key Civil Space History Studies

Atkinson, Joseph D., Jr., and Jay M. Shafritz. *The Real Stuff: A History of the NASA Astronaut Requirements Program*. New York: Praeger, 1985. This overview of the selection of the first 10 groups of NASA astronauts through 1984 places heavy emphasis on the criteria and the procedures used in selecting astronauts.

Atwill, William D. *Fire and Power: The American Space Program As Postmodern Narrative*. Athens: University of Georgia Press, 1994. A unique contribution, this study comments on the development of space exploration as a "big science" program that masked the weaknesses of a "bankrupt" American society.

Bainbridge, William Sims. *The Spaceflight Revolution: A Sociological Study*. New York: Wiley-Interscience, 1976. This impressive study of the rise of space exploration out of its national security component posits an essential conspiracy of spaceflight enthusiasts to persuade public officials that human exploration of the planets was the logical outgrowth of the military effort.

Beattie, Donald A. *Taking Science to the Moon: Lunar Experiments and the Apollo Program*. Baltimore, MD: Johns Hopkins University Press, 2001. This is a fine account of lunar science during the Apollo years, written by a participant.

Benson, Charles D., and William Barnaby Faherty. *Moonport: A History of Apollo Launch Facilities and Operations*. Washington, DC: NASA SP-4204, 1978. This is an excellent history of the design and construction of the lunar launch facilities at Kennedy Space Center.

Bilstein, Roger E. *The American Aerospace Industry: From Workshop to Global Enterprise*. New York: Twayne Publishers, 1996. Reprinted as *The Enterprise of Flight: The American Aviation and Aerospace Industry*. Washington, DC: Smithsonian Institution Press, 2001, Smithsonian History of Aviation and Spaceflight Series. This outstanding overview of the history of this critical industry gives due attention to the spaceflight aspects of its development.

—. *Flight in America: From the Wrights to the Astronauts*. Baltimore, MD: Johns Hopkins University Press, 1984, paperback reprint 1994. A superb synthesis of the origins and development of aerospace activities in America, this is the book to start with in any investigation of air and space activities.

—. *Orders of Magnitude: A History of the NACA and NASA, 1915–1990*. Washington, DC: NASA SP-4406, 1989. This is a very fine nonscholarly general history of the National Aeronautics and Space Administration and its predecessor, the National Advisory Committee for Aeronautics.

—. *Stages to Saturn: A Technological History of the Apollo/Saturn Launch Vehicles*. Washington, DC: NASA SP-4206, 1980, rep. ed. 1997. This thorough and well-written book gives a detailed but highly readable account of the enormously complex process whereby NASA developed the launch vehicles used in the Apollo program ultimately to send 12 humans to the Moon.

Bradbury, Ray, Arthur C. Clarke, Bruce C. Murray, and Carl Sagan. *Mars and the Mind of Man*. New York: Harper and Row, 1973. A superb analysis by a stellar collection of authors, this book discusses the place of the planet Mars in the mythology and science of humanity from the ancients to the late 20th century.

Bromberg, Joan Lisa. *NASA and the Space Industry*. Baltimore, MD: Johns Hopkins University Press, 1999, New Series in NASA History. This is an outstanding history of the interrelationships between NASA and its contractors.

Brooks, Courtney G., James M. Grimwood, and Loyd S. Swenson, Jr. *Chariots for Apollo: A History of Manned Lunar Spacecraft*. Washington: NASA SP-4205, 1979. Based on exhaustive documentary and secondary research as well as 341 interviews, this well-written volume covers the design, development, testing, evaluation, and operational use of the Apollo spacecraft through July 1969.

Bulkeley, Rip. *The Sputniks Crisis and Early United States Space Policy: A Critique of the Historiography of Space*. Bloomington: Indiana University Press, 1991. An important discussion of early efforts to develop civil space policy in the aftermath of the Sputnik crisis of 1957, it contains much information relative to the rivalry between the United States and the Soviet Union and how it was affected by the launching of the Sputnik scientific satellite.

Burrough, Bryan. *Dragonfly: NASA and the Crisis Aboard the Mir*. New York: Ballinger Pub. Co., 1998. Written by the coauthor of the best-selling *Barbarians at the Gate*, this book provides a journalistic analysis of the American-Russian cooperation in space in the mid-1990s aboard the Mir space station. It is culled from one-on-one interviews and transcripts of recorded conversations between the astronauts and cosmonauts on Mir and Russian Mission Control. Burrough delves deeply into the personal and professional lives of the 11 people who lived aboard Mir from 1995 to 1998.

Burrows, William E. *Deep Black: Space Espionage and National Security*. New York: Random House, 1987. This book provides a well-written and thoughtful discussion of the origin and evolution of the United States's satellite reconnaissance program from the 1950s to the late 1980s.

—. *Exploring Space: Voyages in the Solar System and Beyond*. New York: Random House, 1990. Burrows delivers a very well-written and insightful discussion of the robotic probes sent to the planets and what scientists learned from their encounters.

—. *The Infinite Journey: Eyewitness Accounts of NASA and the Age of Space*. New York: Discovery Books, 2000. This large-format history of spaceflight includes important discussions of the endeavor by participants.

—. *This New Ocean: The Story of the First Space Age*. New York: Random House, 1998. Burrows again strikes the right chord in this strong overview of the history of the Space Age from Sputnik to 1998.

Butrica, Andrew J., editor. *Beyond the Ionosphere: Fifty Years of Satellite Communication*. Washington, DC: NASA SP-4217, 1997. This book comprises a useful collection of papers originally offered at a conference about the subject includes essays on international developments.

Byerly, Radford, Jr., editor. *Space Policy Alternatives*. Boulder, CO: Westview Press, 1993. This is an exceptional collection of articles on broad questions in space exploration ranging from the militarization of spaceflight to the place of NASA in the larger schema.

—, editor. *Space Policy Reconsidered*. Boulder, CO: Westview Press, 1989. This is another fine collection of articles on broad questions in space exploration ranging from the militarization of spaceflight to the place of NASA in the larger schema.

Byrnes, Mark E. *Politics and Space: Image Making by NASA*. New York, Praeger, 1994. Byrnes provides reasoned analysis of the approach taken by the U.S. space agency toward developing its own public image.

Caiden, Martin, and Jay Barbree, with Susan Wright. *Destination Mars: In Art, Myth, and Science*. New York: Penguin Studio, 1997. This work is a beautifully illustrated overview of the lure of the Red Planet throughout humanity's history.

Chaikin, Andrew. *A Man on the Moon: The Voyages of the Apollo Astronauts*. New York: Viking, 1994. One of the best books on Apollo, this work emphasizes the exploration of the Moon by the astronauts between 1968 and 1972.

Chaisson, Eric J. *The Hubble Wars: Astrophysics Meets Astropolitics in the Two-Billion-Dollar Struggle over the Hubble Space Telescope*. New York: Harper-Collins, 1994. Chaisson's work is a provocative but not always reliable discussion of the inner workings and conflicts between scientists, engineers, technological managers, the keepers of space science's image, and public policy advocates concerning the Hubble Space Telescope.

Chang, Iris. *Thread of the Silkworm*. New York: Basic Books, 1996. This is the biography of H. S. Tsien, the brilliant Chinese rocketeer who studied at the California Institute of Technology (Caltech) in the 1930s and was deported to the People's Republic China in 1950 as a part of the McCarthy purges. Once there, after being jailed for a time, Tsien became the architect of the Chinese ballistic missile program, developing the Silkworm and other rockets for the military.

Coles, Peter, and Francesco Lucchin. *Cosmology: The Origin and Evolution of Cosmic Structure*. New York: John Wiley, 1995. This book uses knowledge gained from the space exploration program.

Compton, W. David, and Charles D. Benson. *Living and Working in Space: A History of Skylab*. Washington, DC: NASA SP-4208, 1983. This book is the official NASA history of Skylab, an orbital workshop placed in orbit in the early 1970s.

—. *Where No Man Has Gone Before: A History of Apollo Lunar Exploration Missions*. Washington, DC: NASA SP-4214, 1989. This clearly written account traces the Moon landings from the circumlunar mission of Apollo 8 in December 1968 until the final landing of Apollo 17 in December 1972.

Cortright, Edgar M., editor. *Apollo Expeditions to the Moon*. Washington, DC: NASA SP-350, 1975. This large-format volume, with numerous illustrations in both color and black-and-white, contains essays by numerous NASA luminaries.

Darling, David. *Life Everywhere: The Maverick Science of Astrobiology*. New York: Basic Books, 2001. A study of life in the universe is provided in this work.

Dawson, Virginia P. *Engines and Innovation: Lewis Laboratory and American Propulsion Technology*. Washington, DC: NASA SP-4306, 1991. Dawson has written a fine institutional history of the Lewis Research Center from its creation in 1941 to the early 1990s.

Day, Dwayne A., John M. Logsdon, and Brian Latell, editors. *Eye in the Sky: The Story of the Corona Spy Satellite*. Washington, DC: Smithsonian Institution Press, 1998. This work provides a historical overview of the first satellite reconnaissance effort in the United States. Developed in the 1950s and finally successful in launching into orbit in 1960, Corona was operational until 1973.

Dethloff, Henry C. *"Suddenly Tomorrow Came...": A History of the Johnson Space Center*. Washington, DC: NASA SP-4307, 1993. This is the official history of the Manned Spacecraft Center, renamed the Johnson Space Center in 1973, in Houston, Texas, the home of the astronauts and Mission Control.

DeVorkin, David H. *Science with a Vengeance: How the Military Created the US Space Sciences After World War II*. New York: Springer-Verlag, 1992. An excellent analysis of military experiments with rocketry in the immediate postwar era; scientists were able to place experiments on these vehicles, so the flights achieved more than one purpose.

Dick, Steven J. *Plurality of Worlds: The Origins of the Extraterrestrial Life Debate from Democritus to Kant*. Cambridge, England: Cambridge University Press, 1982. A former NASA Chief Historian provides an outstanding discussion of the evolution of humanity's belief in the possibility of life beyond Earth.

—. *The Biological Universe: The Twentieth Century Extraterrestrial Life Debate and the Limits of Science*. New York: Cambridge University Press, 1996. This is the superb continuation of the author's earlier studies on plurality of worlds.

Dickson, Paul. *Sputnik: The Shock of the Century*. New York: Walker and Co., 2001. A fine study of this complex subject is provided.

Divine, Robert A. *The Sputnik Challenge: Eisenhower's Response to the Soviet Satellite*. New York: Oxford University Press, 1993. This book contains insights into the space program as promoted by the Eisenhower White House.

Dorsey, Gary. *Silicon Sky: How One Small Start-Up Went over the Top To Beat the Big Boys into Satellite Heaven*. Reading, MA: Perseus Books, 1999. This book discusses the development of Orbital Sciences Corporation, builder of low-cost satellites and launch vehicles, and gives readers an insider's look at one portion of the life of this aerospace startup. Dorsey focuses on one chapter of Orbital's history, after the company had established itself with the first flights of the Pegasus launcher and was starting work on Orbcomm, a series of low-Earth orbit communications satellites. He then carries the story to 1995. It is a compelling celebration that may provide lessons to the next generation of aerospace startups.

Doyle, Stephen E., editor. *History of Liquid Rocket Engine Development in the United States, 1955–1980.* San Diego, CA: Univelt, 1992, AAS History Series. The editor has provided an excellent collection of papers on the development of liquid rocket technology.

Dunar, Andrew J., and Stephen P. Waring. *Power To Explore: A History of the Marshall Space Flight Center.* Washington, DC: NASA SP-4313, 1999. A fine institutional history of the rocket development center once led by Wernher von Braun.

Emme, Eugene M. *A History of Space Flight*. New York: Holt, Rinehart and Winston, 1965. This short history of space exploration with emphasis on the U.S. human effort was written by the NASA historian in the 1960s.

—, editor. *The History of Rocket Technology: Essays on Research, Development, and Utility*. Detroit, MI: Wayne State University Press, 1964. This is nearly the only work of detail that surveys the sweep of rocket technology development in the United States. Although outdated because of events since publication, the work is unique because it brings together engineers, scientists, and historians to discuss rocketry's development.

Ezell, Edward Clinton, and Linda Neuman Ezell. *On Mars: Exploration of the Red Planet, 1958–1978.* Washington, DC: NASA SP-4212, 1984. This detailed study of NASA's efforts to send space probes to Mars culminates with the soft landing of the two Viking spacecraft in the mid-1970s.

—. *The Partnership: A History of the Apollo-Soyuz Test Project.* Washington, DC: NASA SP-4209, 1978. This is an outstanding, detailed study of the effort by the United States and the Soviet Union in the mid-1970s to conduct a joint human spaceflight.

Fischer, Daniel. *Mission Jupiter: The Spectacular Journey of the Galileo Spacecraft.* New York: Copernicus Books, 2001. Fischer provides a history of the Galileo spacecraft to Jupiter.

Fitzgerald, Frances. *Way Out There in the Blue: Reagan, Star Wars, and the End of the Cold War*. New York: Simon & Schuster, 2000. The author provides an important history of the Strategic Defense Initiative.

Fleming, James Rodger. *Historical Perspectives on Climate Change.* New York: Oxford University Press, 1998. Fleming writes a solid effort to understand the history of Earth science.

French, Bevan M., and Stephen P. Maran, editors. *A Meeting with the Universe: Science Discoveries from the Space Program*. Washington, DC: NASA Educational Publication-177, 1981. This is a very fine discussion of the scientific results of NASA's efforts to explore the planets of the solar system and the wider universe.

Friedman, Norman. *Seapower and Space: From the Dawn of the Missile Age to Net-Centric Warfare*. Annapolis, MD: Naval Institute Press, 2000. Friedman provides an interesting study of historical analogy.

Fries, Sylvia D. *NASA Engineers and the Age of Apollo*. Washington, DC: NASA SP-4104, 1992. This book is a sociocultural analysis of a selection of engineers at NASA who worked on Project Apollo. It analyzes the manner in which different personalities, perspectives, backgrounds, and priorities came together to inform the direction of NASA during the 1960s.

Gall, Sarah L., and Joseph T. Pramberger. *NASA Spinoffs: 30-Year Commemorative Edition*. Washington, DC: NASA, 1992. This work is a better-than-average recitation of the technologies originally developed to support space exploration and their transfer to Earth-based commercial uses.

Gavaghan, Helen. *Something New Under the Sun: Satellites and the Beginning of the Space Age*. New York: Copernicus Books, 1998. This is the first book focusing on the history and development of satellite technology. Highlighting three major areas of development—navigational satellites, communications, and weather observation and forecasting—this book tells the remarkable inside story of how obscure men and women, often laboring under strict secrecy, made satellite technology possible.

Glennan, T. Keith. *The Birth of NASA: The Diary of T. Keith Glennan*. Edited by J. D. Hunley. Washington, DC: NASA SP-4105, 1993. This diary of NASA's first administrator contains a detailed account of the rise and development of the Agency between 1958 and the end of 1960.

Goddard, Esther C., editor, and G. Edward Pendray, associate editor. *The Papers of Robert H. Goddard*. 3 vols. New York: McGraw-Hill Book Co., 1970. This is an excellent collection of primary source materials that describe the life and work of the premier American rocketry experimenter of the first half of the 20th century.

Goldsmith, Donald. *The Hunt for Life on Mars*. New York: E. P. Dutton, 1997. This work is a solid discussion of the study of Mars, especially the search for possible life there, from Percival Lowell's work in the late 19th century to present-day efforts.

Goldstein, Stanley H. *Reaching for the Stars: The Story of Astronaut Training and the Lunar Landing*. New York: Praeger, 1987. This is a detailed account of the development and management of the astronaut training program for Project Apollo.

Gorn, Michael H. *The Universal Man: Theodore von Kármán's Life in Aeronautics*. Washington, DC: Smithsonian Institution Press, 1992. Gorn has written a fine biography of the originator and first head of the organization that eventually became the Jet Propulsion Laboratory in Pasadena, California.

Gray, Mike. *Angle of Attack: Harrison Storms and the Race to the Moon.* New York: W.W. Norton and Co., 1992. This is a lively journalistic account of the career of Harrison Storms, president of the Aerospace Division of North American Aviation, which built the Apollo capsule.

Greeley, Ronald, and Raymond Batson. *The NASA Atlas of the Solar System.* New York: Cambridge University Press, 1996. This landmark book is on space science as understood at the end of the 20th century.

Green, Constance M., and Milton Lomask. *Vanguard: A History.* Washington, DC: NASA SP-4202, 1970; rep. ed. Smithsonian Institution Press, 1971. Green provides an excellent account of the development and operation of what was supposed to be the United States' first orbital satellite in the 1950s.

Griffith, Alison. *The National Aeronautics and Space Act: A Study of the Development of Public Policy.* Washington, DC: Public Affairs Press, 1962. This is a detailed study of the process of writing this major legislation in 1958.

Grinspoon, David Harry. *Venus Revealed: A New Look Below the Clouds of Our Mysterious Twin Planet.* Reading, MA: Addison-Wesley Pub. Co., Helix Books, 1997. Grinspoon writes an account of the Magellan radar mapping mission to Venus in 1989–90.

Hacker, Barton C., and James M. Grimwood. *On Shoulders of Titans: A History of Project Gemini.* Washington, DC: NASA SP-4203, 1977. This work is the official history of the Gemini project conducted by NASA in the mid-1960s.

Hall, R. Cargill. *Lunar Impact: A History of Project Ranger.* Washington, DC: NASA SP-4210, 1977. The official history of the Ranger program to send robotic probes to the Moon in the late 1950s and 1960s.

—, and Jacob Neufeld, editors. *The U.S. Air Force in Space: 1945 to the 21st Century.* Washington, DC: USAF History and Museums Program, 1998. This collection of essays explores the range of activities in space undertaken by the Air Force.

Hallion, Richard P. *On the Frontier: Flight Research at Dryden, 1946–1981.* Washington, DC: NASA SP-4303, 1984. This institutional history of Dryden Flight Research Center (now known as Armstrong Flight Research Center) discusses the NASA facility in the Mojave Desert where hypersonic vehicles such as the X-15 were flown.

—, and Tom D. Crouch, editors. *Apollo: Ten Years Since Tranquility Base.* Washington, DC: Smithsonian Institution Press, 1979. This is a collection of 16 essays developed for the National Air and Space Museum to commemorate the 10th anniversary of the first landing on the Moon.

Handberg, Roger. *Seeking New World Vistas: The Militarization of Space.* Westport: Praeger Publishers, 2000. An interesting analysis of military space policy.

—, and Joan Johnson-Freese. *The Prestige Trap: A Comparative Study of the U.S., European, and Japanese Space Programs*. Dubuque, IA: Kendall/Hunt Publishing Co., 1994. An interesting study of the various programs and their development over the years.

Hanle, Paul A., and Del Chamberlain, editors. *Space Science Comes of Age: Perspectives in the History of the Space Sciences*. Washington, DC: Smithsonian Institution Press, 1981. A superior collection of essays on all aspects of the space sciences.

Harford, James J. *Korolev: How One Man Masterminded the Soviet Drive To Beat America to the Moon*. New York: John Wiley & Sons, 1997. The first English-language biography of the Soviet "chief designer" who directed the projects that were so successful in the late 1950s and early 1960s in energizing the Cold War rivalry for space supremacy.

Harland, David M. *Exploring the Moon: The Apollo Expeditions*. Chichester, U.K.: Wiley-Praxis, 1999. This work focuses on the exploration and science mission carried out by Apollo astronauts while on the lunar surface.

—. *Jupiter Odyssey: The Story of NASA's Galileo Mission*. Chichester, U.K.: Springer-Praxis, 2000, Springer-Praxis Books in Astronomy and Space Sciences. This is the first book to offer a history of the Galileo space probe sent to Jupiter.

—. *The Space Shuttle: Roles, Missions and Accomplishments*. Chichester, U.K.: Wiley-Praxis, 1998. This book is the most sophisticated of any Shuttle history to appear to date.

Harvey, Dodd L., and Linda C. Ciccoritti. *U.S.-Soviet Cooperation in Space*. Miami, FL: Center for Advanced International Studies at the University of Miami, 1974, Monographs in International Affairs. This is a detailed exploration of the competition/cooperation in space exploration by the two superpowers of the Cold War era through the détente that led to the joint Apollo-Soyuz Test Project.

Hayes, Peter L., James M. Smith, Alan R. Van Tassel, and Guy M. Walsh, editors. *Spacepower for a New Millennium: Space and U.S. National Security*. New York: McGraw Hill for the United States Air Force Institute for National Security Studies, 2000. This is a space policy analysis.

Hechler, Ken. *Toward the Endless Frontier: History of the Committee on Science and Technology, 1959–1979*. Washington, DC: U.S. House of Representatives, 1980. This work contains the best account to date of congressional wrangling over space exploration and demonstrates the bipartisan nature of both its support and its opposition.

Henbest, Nigel, and Heather Couper. *The Guide to the Galaxy*. New York: Cambridge University Press, 1994. This interesting basic handbook examines the present scientific understanding of the Milky Way Galaxy.

Heppenheimer, T. A. *Countdown: The History of Space Exploration*. New York: John Wiley & Sons, 1997. This general-audience history is somewhat quirky but well written and entertaining.

Herring, Mack R. *Way Station to Space: A History of the John C. Stennis Space Center*. Washington, DC: NASA SP-4310, 1997. Herring presents the official history of the rocket-testing center built for Project Apollo.

Hickam, Homer H., Jr. *Rocket Boys: A Memoir*. New York: Delacorte Press, 1998. In the fall of 1957, the launch of Sputnik electrified the United States. That was true even in the small, isolated town of Coalwood, West Virginia, where outside events were rarely favored as topics of discussion over local gossip. It also grabbed the intention of Homer Hickam, Jr., younger son of the local coal mine supervisor, who announced out of the blue his intention to build a rocket. Homer's efforts to build rockets, and the effect on his community, are the subject of this memoir.

Hudson, Heather E. *Communications Satellites: Their Development and Impact*. New York: Free Press, 1990. No space technology has held more importance for modern America than the communications satellite, and this book attempts to discuss this application and how it has changed our lives.

Hufbauer, Karl. *Exploring the Sun: Solar Science Since Galileo*. Baltimore, MD: Johns Hopkins University Press, 1991. This prize-winning history of the development of solar science since the 15th century places emphasis on the 20th-century contribution made possible because of the advent of the Space Age.

Hujsak, Edward. *The Future of U.S. Rocketry*. La Jolla, CA: Mina-Helwig Company, 1994. More a history than anything else, this book describes the development of various types of launch vehicles.

Jenkins, Dennis R. *Space Shuttle: The History of the National Space Transportation System, the First 100 Missions*. Cape Canaveral, FL: Dennis R. Jenkins, 2001, 4th ed. Perhaps the best technical history, this book presents an overview of the Space Shuttle and its development and use.

Johnson, Dana J., Scott Pace, and C. Bryan Gabbard. *Space: Emerging Options for National Power*. Santa Monica, CA: RAND, 1998. With the Cold War over and with growing commercial use of space by the United States and other countries, the military's role in space needs to be reevaluated. The authors outline the shifts in expected military operations in a post–Cold War world and provide three options for future military use of space. A "minimalist" option would have the military rely on partnerships with other government agencies and commercial resources to serve the military. An "enhanced" option would expand the military's use of space in a variety of areas and integrate them into the existing military branches. The third option would establish an "aerospace force" separate from the other military branches that would work jointly with other branches but also be able to exercise power separately. The book does not provide a recommendation for any of the three options, choosing instead to characterize the strengths and weaknesses of each.

Johnson-Freese, Joan. *Changing Patterns of International Cooperation in Space*. Malabar, FL: Orbit Books, 1990. Johnson-Freese presents an interesting exploration of the movement from competition to cooperation in space exploration.

—, and Roger Handberg. *Space: The Dormant Frontier, Changing the Paradigm for the 21st Century*. Westport, CT: Praeger, 1997. This political science study of the evolution of space policy since Sputnik offers suggestions for future efforts.

Kauffman, James L. *Selling Outer Space: Kennedy, the Media, and Funding for Project Apollo, 1961–1963*. Tuscaloosa: University of Alabama Press, 1994. This straightforward but helpful history examines the public image–building efforts of NASA and the relation of that image to public policy.

Kay, W. D. *Can Democracies Fly in Space? The Challenge of Revitalizing the U.S. Space Program*. Westport, CT: Praeger, 1995. The answer to the question, of course, is that they can and do, but the analysis here suggests that the political requirement to build broad constituencies and to water down proposals to ensure success, as well as to make claims that are beyond reach, ensure that the effort is inefficient and incremental and always shifting rather than focused and accepting of returns that will accrue beyond the next election.

Koerner, David, and Simon LeVay. *Here There Be Dragons: The Scientific Quest for Extraterrestrial Life*. New York: Oxford University Press, 2000. A good overview.

Koppes, Clayton R. *JPL and the American Space Program: A History of the Jet Propulsion Laboratory*. New Haven, CT: Yale University Press, 1982. An institutional history of one of NASA's major centers of space science activities.

Kosloski, Lillian D. *U.S. Space Gear: Outfitting the Astronaut*. Washington, DC: Smithsonian Institution Press, 1993. Kosloski's work is the only serious history of spacesuits available.

Kraemer, Robert S. *Beyond the Moon: A Golden Age of Planetary Exploration, 1971–1978*. Washington, DC: Smithsonian Institution Press, 2000. This fine discussion was written by a participant.

Krug, Linda T. *Presidential Perspectives on Space Exploration: Guiding Metaphors from Eisenhower to Bush*. New York: Praeger, 1991. This interesting analysis of spaceflight speeches by the presidents emphasizes their use of romantic analogy and metaphor in their rhetoric about space exploration.

Lambakis, Steven James. *On the Edge of Earth: The Future of American Space Power*. Lexington: University Press of Kentucky, 2001. Lambakis discusses military space policy.

Lambright, W. Henry. *Powering Apollo: James E. Webb of NASA*. Baltimore, MD: Johns Hopkins University Press, 1995. This is an excellent biography of the NASA Administrator between 1961 and 1968, the critical period in which Project Apollo was under way.

Launius, Roger D. *Frontiers of Space Exploration*. Westport, CT: Greenwood Press, 1998, Critical Events in the Twentieth Century Series. This book presents a collection of essays along with key documents and biographies of actors in the space exploration effort.

—, and Howard E. McCurdy. *Imagining Space: Achievements, Predictions, Possibilities, 1950–2050*. San Francisco, CA: Chronicle Books, 2001. This large-format book sweeps from the past to the future of spaceflight.

—, editors. *Spaceflight and the Myth of Presidential Leadership*. Urbana: University of Illinois Press, 1997. This collection of essays covers Presidents Eisenhower, Kennedy, Johnson, Nixon, Ford, and Carter, with additional discussions of international cooperation and the role of the presidency in shaping space policy.

Launius, Roger D. With Bertram Ulrich. *NASA & the Exploration of Space*. New York: Stewart, Tabori, and Chang, 1998. This basic history of NASA is illustrated with works from the NASA art program.

—. *NASA: A History of the U.S. Civil Space Program*. Melbourne, FL: Krieger Pub. Co., 1994. A short book in the Anvil Series, this history of U.S. civilian space efforts consists half of narrative and half of documents.

—, editor. *Organizing for the Use of Space: Historical Perspectives on a Persistent Issue*. San Diego, CA: Univelt, Inc., 1995, AAS History Series, Vol. 18. This collection of essays related to the organizational elements of conducting operations in space is organized chronologically and focuses on both civil and military aspects of the U.S. effort.

—, John M. Logsdon, and Robert W. Smith, editors. *Reconsidering Sputnik: Forty Years Since the Soviet Satellite*. Amsterdam, The Netherlands: Harwood Academic Publishers, 2000. This book comprises a collection of essays.

Lemonick, Michael D. *Other Worlds: The Search for Life in the Universe*. New York: Simon & Schuster, 1998. This is a strong analysis of whether or not there is life on other planets in the universe and efforts under way to find it.

Leverington, David. *New Cosmic Horizons: Space Astronomy from the V-2 to the Hubble Space Telescope*. New York: Cambridge University Press, 2001. Leverington's book is an outstanding history of this subject.

Levine, Alan J. *The Missile and Space Race*. New York: Praeger, 1994. A somewhat quirky work, this study presents some interesting perspectives on the development of the rivalry between the United States and the Soviet Union in space exploration.

Levine, Arnold S. *Managing NASA in the Apollo Era*. Washington, DC: NASA SP-4102, 1982. A narrative account of NASA from its origins through 1969, this book analyzes key administrative decisions, contracting, personnel, the budgetary process, headquarters organization, relations with the Department of Defense, and long-range planning.

Lewis, John S. *Worlds Without End: The Exploration of Planets Known and Unknown.* Reading, MA: Perseus Books, 1998. This book is a history of the search for planets beyond this solar system.

—, and Ruth A. Lewis. *Space Resources: Breaking the Bonds of Earth.* New York: Columbia University Press, 1987. An account of the possibilities for commercial enterprises in space.

Ley, Willy. *Rockets, Missiles, and Men in Space.* New York: Viking Press, 1968. This is the fourth and final edition of 21 printings of the work first published as *Rockets.* It emphasizes the possibilities of spaceflight as a reality rather than science fiction. Ley came to the United States in 1935, and this book became one of the most significant textbooks available in the mid-20th century on the possibilities of space travel. Once again, the book emphasizes the importance of a trip to the Moon as humanity's first step off Earth and into the universe.

Light, Michael. *Full Moon.* New York: Alfred A. Knopf, 1999. In this book, Michael Light has woven 129 of these stunningly clear images into a single composite voyage, a narrative of breathtaking immediacy and authenticity.

Link, Mae Mills. *Space Medicine in Project Mercury.* Washington, DC: NASA SP-4003, 1965. The first program to launch an American into space was enormously important from the perspective of biomedicine. Could humans survive the rigors of launch on rockets and the harshness of the vacuum of space? This book surveys the questions asked by scientists prior to the project's first launch and what answers the experiments with the first astronauts yielded in terms of information on this subject.

Logsdon, John M. *The Decision To Go to the Moon: Project Apollo and the National Interest.* Cambridge, MA: MIT Press, 1970. This classic study analyzes the political process in the United States leading to the decision to go to the Moon in 1961.

—, general editor. *Exploring the Unknown: Selected Documents in the History of the U.S. Civil Space Program.* 6 vols. Washington, DC: NASA SP-4407, 1995–2008. An essential reference work, these volumes contain more than 650 key documents about space policy and its development throughout the 20th century.

Mack, Pamela E. *Viewing the Earth: The Social Construction of Landsat.* Cambridge, MA: MIT Press, 1990. Mack's book is the only substantive study of the origins and development of the first Earth resources–monitoring satellite program, launched in the 1970s.

Mailer, Norman. *Of a Fire on the Moon.* Boston: Little, Brown, 1970; London: Weidenfeld & Nicolson, 1970; New York: New American Library, 1971. Written by one of the foremost contemporary American writers, this book shines as an exploration of how the 1960s' countercultural mindset meets its antithesis, a NASA steeped in middle-class values and reverence for the American flag and culture.

Mather, John, and John Boslough. *The Very First Light: The True Inside Story of the Scientific Journey Back to the Dawn of the Universe.* New York: Basic Books, 1996. A solid account of NASA's Cosmic Background Explorer written by the project's chief scientist.

McCurdy, Howard E. *Faster, Better, Cheaper: Low-Cost Innovation in the U.S. Space Program.* Baltimore, MD: Johns Hopkins University Press, 2001, New Series in NASA History. This book studies the history of NASA's efforts to reform itself in the 1990s.

—. *Inside NASA: High Technology and Organizational Change in the U.S. Space Program.* Baltimore, MD: Johns Hopkins University Press, 1993. This work discusses the evolution of the NASA organizational culture from the creation of the Agency to the 1990s using extensive interviews with key personnel and documentary sources.

—. *Space and the American Imagination.* Washington, DC: Smithsonian Institution Press, 1997. This book is a significant analysis of the relationship between popular culture and public policy.

—. *The Space Station Decision: Incremental Politics and Technological Choice.* Baltimore, MD: Johns Hopkins University Press, 1990. This work is a fine study of the political process that led to the presidential decision in 1984 to develop an orbital space station.

McDonald, Robert A. *Corona Between the Sun and the Earth: The First NRO Reconnaissance Eye in Space.* Bethesda, MD: ASPRS Publications, 1997. McDonald presents a historical overview of the first satellite reconnaissance effort in the United States. Developed in the 1950s and finally successful in launching into orbit in 1960, Corona was operational until 1973.

McDougall, Walter A. *...The Heavens and the Earth: A Political History of the Space Age.* New York: Basic Books, 1985; rep. ed. Baltimore, MD: Johns Hopkins University Press, 1997. This Pulitzer Prize–winning book analyzes the space race to the Moon in the 1960s. The author argues that Apollo prompted the space program to stress engineering over science, competition over cooperation, civilian over military management, and international prestige over practical applications.

McLucas, John L. *Space Commerce.* Cambridge, MA: Harvard University Press, 1991. This book is a useful short analysis of the rise and evolution of business opportunities in space, among them communications satellite technology.

Michaud, Michael A. G. *Reaching for the High Frontier: The American Pro-Space Movement, 1972–1984.* New York: Praeger, 1986. Michaud presents a cogent history and commentary of the pro-space efforts made by voluntary organizations that arose near the end of the Apollo program, identifying key groups, tracing their origins and goals, and describing how they had a subtle but critical influence on the space policy of the nation during the formative years of Shuttle development.

Morrison, David. *Exploring Planetary Worlds*. New York: Scientific American Books, 1993. Using data provided from the space probes sent to the planets, this work is another fine discussion of the planets and other bodies of the solar system.

Muenger, Elizabeth A. *Searching the Horizon: A History of Ames Research Center, 1940–1976*. Washington, DC: NASA SP-4304, 1985. Muenger presents an institutional history of the NASA Center charged with researching aerospace medicine as well as other problems of spaceflight.

Murray, Bruce C. *Journey into Space: The First Three Decades of Space Exploration*. New York: W.W. Norton and Co., 1989. This book, written by a former Director of the Jet Propulsion Laboratory, is an excellent discussion of the planetary science program.

Murray, Charles A., and Catherine Bly Cox. *Apollo: The Race to the Moon*. New York: Simon & Schuster, 1989; rep. ed. Burkittsville, MD: South Mountain Books, 2004. This is perhaps the best general account of the lunar program, this history uses interviews and documents to reconstruct the stories of the people who participated in Apollo.

NASA Historical Data Book. 4 vols. Washington, DC: NASA SP-4012, 1976–1994. Presented here is an exhaustive compendium of basic information on NASA resources, programs, and projects during the period between 1958 and 1978.

Nash, Philip. *The Other Missiles of October: Eisenhower, Kennedy, and the Jupiters, 1957–1963*. Chapel Hill: University of North Carolina Press, 1997. This book is a fine, detailed study of one case in which ballistic missile technology entered the sphere of international relations in the Cold War.

Naugle, John E. *First Among Equals: The Selection of NASA Space Science Experiments*. Washington, DC: NASA SP-4215, 1991. This work is an outstanding discussion of the origin and evolution of NASA's space science experiment decision-making process.

Neal, Valerie, Cathleen S. Lewis, and Frank H. Winter. *Spaceflight: A Smithsonian Guide*. New York: Macmillan, 1995. This book provides, with numerous illustrations, a basic history of space exploration by the United States.

—, editor. *Where Next, Columbus? The Future of Space Exploration*. New York: Oxford University Press, 1994. Neal provides an excellent collection of essays linking the voyage of discovery by Columbus with human exploration of space.

Needell, Allan A., editor. *The First 25 Years in Space: A Symposium*. Washington, DC: Smithsonian Institution Press, 1983. This book is a collection of papers from a conference on space exploration held in commemoration of the 1957 launching of Sputnik.

Neufeld, Jacob. *Ballistic Missiles in the United States Air Force, 1945–1960.* Washington, DC: Center for Air Force History, 1990. This work is an official history of the Air Force's efforts to develop ballistic missiles, many of which also later served as launch vehicles for space probes and human flight.

Neufeld, Michael J. *The Rocket and the Reich: Peenemünde and the Coming of the Ballistic Missile Era.* New York: The Free Press, 1995. Neufeld presents the finest study yet of the German effort in World War II to develop the V-2 ballistic missile.

—. *Wernher von Braun.* New York: Alfred A. Knopf, 2007. This book is the indispensable biography of the father of the German V-2, the Jupiter ballistic missile, and the Saturn V Moon rocket.

Newell, Homer E. *Beyond the Atmosphere: Early Years of Space Science.* Washington, DC: NASA SP-4211, 1980. The Agency's first chief scientist provides a thoughtful and revealing memoir of space science in NASA during the 1950s and 1960s.

Nicks, Orin W. *The Far Travelers: The Exploring Machines.* Washington, DC: NASA SP-480, 1985. This work is a somewhat idiosyncratic discussion of the development of space probes at NASA for the conduct of space science from the beginning of the Space Age through the early 1980s.

—, editor. *This Island Earth.* Washington, DC: NASA SP-250, 1970. Something of a classic in the history of government printing, this exciting book collects ruminations on Earth as a fragile lifeboat hanging in the harshness of space. It reflects and perhaps contributes to the rise of the modern environmental movement in the United States that was given additional steam by the images of Earth taken from beyond the planet and provided for the first time by the space program.

Ordway, Frederick I., III. *Visions of Spaceflight: Images from the Ordway Collection.* New York: Four Walls Eight Windows, 2001. Ordway presents a cultural history.

—, and Randy Lieberman, editors. *Blueprint for Space: From Science Fiction to Science Fact.* Washington, DC: Smithsonian Institution Press, 1992. This book is a fine collection of essays, accompanied by spectacular artwork and photographs, dealing with the popular culture of spaceflight in the 20th century.

—, and Mitchell R. Sharpe. *The Rocket Team.* New York: Crowell, 1979. This is an important, popularly oriented, and somewhat apologetic discussion of the activities of the group of German engineers under the leadership of Wernher von Braun who developed the V-2 in World War II, came to the United States in 1945, and worked at Marshall Space Flight Center in Huntsville, Alabama, to develop the Saturn V launch vehicle used in Project Apollo.

Peebles, Curtis. *The Corona Project: America's First Spy Satellites.* Annapolis, MD: Naval Institute Press, 1997. This work is a historical overview of the first satellite reconnaissance effort in the United States. Developed in the 1950s and finally successful in launching into orbit in 1960, Corona was operational until 1973.

—. *Watch the Skies! A Chronicle of the Flying Saucer Myth*. Washington, DC: Smithsonian Institution Press, 1994. This is the best historical study to appear on the debate over the possibility of continuing visitation by extraterrestrials. Contains extensive documentation and quotations from official records.

Pellegrino, Charles R., and Joshua Stoff. *Chariots for Apollo: The Making of the Lunar Module*. New York: Atheneum, 1985. This book is a popular but not-always-accurate discussion of the development of the Lunar Module by the Grumman Aerospace Corporation.

Penley, Constance. *NASA/TREK: Popular Science and Sex in America*. New York: Verso, 1997. This work is a truly provocative postmodern analysis of spaceflight and its meaning in the development of modern America.

Pisano, Dominick A., and Cathleen S. Lewis, editors. *Air and Space History: An Annotated Bibliography*. New York: Garland Publishing Co., Inc., 1988. This exhaustive reference is the place to start when searching for bibliographical information about any aspect of aerospace history.

Pitts, John A. *The Human Factor: Biomedicine in the Manned Space Program to 1980*. Washington, DC: NASA SP-4213, 1985. This account traces the history of space medicine from its early days before the founding of NASA through the decade following the Apollo program.

Portree, David S. F. *Humans to Mars: Fifty Years of Mission Planning, 1950–2000*. Washington, DC: NASA SP-2001-4521, 2001. Portree provides an important analysis of a 50-year effort to send humans to the Red Planet.

Redfield, Peter. *Space in the Tropics: From Convicts to Rockets in French Guiana*. Berkeley: University of California, 2000. Redfield has created a terrific discussion of the development of the European spaceport.

Reeves, Robert. *The Superpower Space Race: An Explosive Rivalry Through the Solar System*. New York: Plenum Press, 1994. Somewhat mistitled—suggesting the broadest possible context of discussion for U.S./USSR rivalry in space exploration—this book is a well-done journalistic account of the robotic race to the various planets by scientists of both nations.

Richelson, Jeffrey T. *America's Secret Eyes in Space: The U.S. Keyhole Spy Satellite Program*. New York: Harper and Row, 1990. This is the best study yet of the U.S. military reconnaissance satellite program from its inception in the 1950s through the end of the Cold War.

—, project director. *U.S. Military Uses of Space, 1945–1991*. Alexandria, VA: Chadwyck-Healey, 1991. Guide and Microfilm Collection. This is an important microfiche collection of formerly classified materials obtained from the Department of Defense and relative to military operations in space.

Roman, Peter J. *Eisenhower and the Missile Gap*. Ithaca, NY: Cornell University Press, 1995. Using recently declassified records, Roman provides an excellent modern analysis of the "missile gap" controversy that arose after the Soviet launch of Sputnik.

Rosenthal, Alfred. *Venture into Space: Early Years of Goddard Space Flight Center*. Washington, DC: NASA SP-4301, 1968. This book is an institutional history of the NASA Center created in 1960 and charged with studying Earth science and aspects of planetary sciences.

Rosholt, Robert L. *An Administrative History of NASA, 1958–1963*. Washington, DC: NASA SP-4101, 1966. This is a particularly valuable work for its discussion of the details of the organization, structure, budget, and procedures of NASA during its earliest years.

Roth, Ladislav E., and Stephen D. Page. *The Face of Venus: The Magellan Radar-Mapping Mission*. Washington, DC: NASA SP-520, 1995. This is a large-format, well-illustrated discussion of the radar-mapping mission conducted by NASA's Magellan spacecraft on its mission to Venus in 1989–90.

Sagan, Carl. *Broca's Brain: Reflections on the Romance of Science*. New York: Ballantine Books, 1974. Written by the premier American space scientist of the latter 20th century, this wide-ranging collection of essays explores issues of life beyond Earth in the universe and the possibilities of human exploration of the cosmos.

—. *Cosmos*. New York: Random House, 1980. Accompanying a stunning PBS science series on the evolution of the universe, this well-illustrated book makes scientific ideas—many coming from research undertaken as part of the space program—both comprehensible and exciting. Despite its age, this is still an outstanding starting point for any research into our universe's origin and evolution.

—. *Pale Blue Dot: A Vision of the Human Future in Space*. New York: Random House, 1994. This work is probably the most sophisticated articulation of the exploration imperative to appear since Wernher von Braun's work of the 1950s and 1960s.

—, Frank D. Drake, Ann Druyan, Timothy Ferris, Jon Lomberg, and Linda Salzman Sagan. *Murmurs of Earth: The Voyager Interstellar Record*. New York: Random House, 1978. This book presents a discussion of conceptualization and carrying out of the effort to place a gold record on the two Voyager spacecraft sent outside the solar system in the 1970s. This record contained digital information on the planet Earth, including photographs, sounds, music, and greetings in more than 40 languages. It was designed to tell an extraterrestrial intelligence who encountered it something about this planet and the life that thrives here, and to give that life-form a general idea of where Earth was located in space.

—, and Thornton Page, editors. *UFOs: A Scientific Debate*. New York: W.W. Norton and Co., 1973. In this outstanding collection of essays relating to the possibility that there is life beyond Earth in the universe and that intelligent forms of life are visiting Earth, the papers suggest that there is no real proof of the existence of extraterrestrial life as yet, although many people believe it must be out there somewhere.

Shapland, David, and Michael Rycroft. *Spacelab: Research in Earth Orbit*. Cambridge, U.K.: Cambridge University Press, 1984. Shapland provides a useful discussion of the development and flight of the laboratory built by Europeans for use aboard the Space Shuttle in Earth orbit.

Shayler, David J. *Disasters and Accidents in Manned Spaceflight*. Chichester, U.K.: Springer-Praxis, 2000, Springer-Praxis Books in Astronomy and Space Sciences. Shayler provides a fine discussion of this subject.

Sheehan, William. *The Planet Mars: A History of Observation & Discovery*. Tucson: University of Arizona Press, 1996. An excellent survey of how humans have acquired knowledge about the Red Planet from antiquity to the present, the book concentrates on the work of Earth-based astronomers but also includes succinct narratives of the Mariner 4 mission and the Viking project of the 1970s.

Siddiqi, Asif A. *Challenge to Apollo: The Soviet Union and the Space Race, 1945–1974*. Washington, DC: NASA SP-2000-4408, 2000. This is the new standard in knowledge about the Soviet side of the race to the Moon.

Sloop, John L. *Liquid Hydrogen as a Propulsion Fuel, 1945–1959*. Washington, DC: NASA SP-4404, 1978. The author gives a detailed discussion of the development of the cryogenic fuels used in rockets.

Smith, Robert W. *The Space Telescope: A Study of NASA, Science, Technology, and Politics*. New York: Cambridge University Press, 1989, rev. ed. 1994. Smith's prize-winning history details the development of the Hubble Space Telescope.

Spires, David N. *Beyond Horizons: A Half Century of Air Force Space Leadership*. Peterson Air Force Base, CO: Air Force Space Command, 1997. Spires provides a fine overview of the subject.

Stares, Paul B. *The Militarization of Space: U.S. Policy, 1945–1984*. Ithaca, NY: Cornell University Press, 1985. This seminal book establishes themes and provides analysis that has not yet been superseded although it is now more than two decades old. Stares suggests that national security needs drove much of what the United States established as its policy toward space exploration.

Stoker, Carol A., and Carter Emmart, editors. *Strategies for Mars: A Guide to Human Exploration*. San Diego, CA: Univelt, Inc., 1996. The most up-to-date and useful of several books related to Mars exploration, this collection of essays provides a rationale, technology assessment, and political analysis of the endeavor through the lens of quite a lot of historical perspective.

Stumpf, David K. *Titan II: A History of a Cold War Missile Program*. Fayetteville: University of Arkansas Press, 2000. Stumpf has written an outstanding study of a single launch vehicle.

Swanson, Glen E., editor. *"Before This Decade Is Out…": Reflections on the Apollo Program*. Washington, DC: NASA SP-4223, 1999. Swanson's work is an outstanding collection of oral histories by some of the key participants.

Swenson, Loyd S., Jr., James M. Grimwood, and Charles C. Alexander. *This New Ocean: A History of Project Mercury*. Washington, DC: NASA SP-4201, 1966. The official history of Project Apollo, this book is based on extensive research and interviews.

Swift, David W. *SETI Pioneers: Scientists Talk About Their Search for Extraterrestrial Intelligence.* Tucson: University of Arizona Press, 1990. No issue has more affected space exploration than the possibility of making contact with life beyond Earth, and this book prints interviews with several scientists and engineers involved in the Search for Extraterrestrial Intelligence, a formal program sponsored by several U.S. organizations.

—. *Voyager Tales: Personal Views of the Grand Tour.* Reston, VA: American Institute of Aeronautics and Astronautics, 1997. These oral histories from the Voyager program emphasize the Jet Propulsion Laboratory experience.

Tatarewicz, Joseph N. *Space Technology and Planetary Astronomy.* Bloomington: Indiana University Press, 1990. Tatarewicz discusses the interrelationships of space scientists, government patrons, and the politics of big science in the period between the 1950s and the 1980s.

Tucker, Wallace H., and Karen Tucker. *Revealing the Universe: The Making of the Chandra X-Ray Observatory.* New York: Harvard University Press, 2001. This work is an important history of a major recent NASA space science effort.

Van Dyke, Vernon. *Pride and Power: The Rationale of the Space Program.* Urbana: University of Illinois Press, 1964. This analysis of the overall rationale of the space program came to the conclusion that the most powerful motives behind it involved competition with the Soviet Union.

Vaughan, Diane. *The Challenger Launch Decision: Risky Technology, Culture, and Deviance at NASA.* Chicago: University of Chicago Press, 1996. The first thorough scholarly study of the events leading to the fateful decision to launch Challenger in January 1986, this book uses sociological and communication theory to piece together the story of America's worst disaster in spaceflight up to that point and to analyze the nature of risk in high-technology enterprises.

Von Benke, Matthew J. *The Politics of Space: A History of U.S.-Soviet/Russian Competition and Cooperation in Space.* Boulder, CO: Westview Press, 1997. This work provides an important analysis of the relationship between the United States and the Soviet Union/Russia since the 1950s.

Von Braun, Wernher, Frederick I. Ordway III, and Dave Dooling. *History of Rocketry and Space Travel.* New York: Thomas Y. Crowell Co., 1986 ed. This is a large-format, illustrated history that emphasizes the history of U.S. space activities. It was written by one of the most significant popularizers of spaceflight with the assistance of professional writers.

Wallace, Harold D., Jr. *Wallops Station and the Creation of the American Space Program.* Washington, DC: NASA SP-4311, 1997. Wallace has written the official history of the small NASA facility where sounding rockets were launched beginning in the latter 1940s.

Webb, James E. *Space Age Management: The Large Scale Approach.* New York: McGraw-Hill Book Co., 1969. Based on a series of lectures, this book by the former NASA Administrator tries to apply the concepts of large-scale technological management employed in Apollo to the other problems of society.

Whalen, David J. *The Origins of Satellite Communications, 1945–1965.* Washington, DC: Smithsonian Institution Press, 2002, Smithsonian History of Aviation and Spaceflight Series. This significant synthesis of the early years of technology development and use was written with emphasis on the role of industry.

Wilford, John Noble. *Mars Beckons: The Mysteries, the Challenges, the Expectations of Our Next Great Adventure in Space.* New York: Alfred A. Knopf, 1990. This superior explanation of the possibilities of Mars exploration includes a discussion of earlier plans to send humans to the Red Planet.

Wilhelms, Don E. *To a Rocky Moon: A Geologist's History of Lunar Exploration.* Tucson: University of Arizona Press, 1993. This work provides a detailed and contextual account of lunar geology during the 1960s and 1970s, as well as a less-detailed but informative account for the rest of the century.

Winter, Frank H. *Prelude to the Space Age: The Rocket Societies, 1924–1940.* Washington, DC: Smithsonian Institution Press, 1983. Winter provides a very good discussion of the private organizations of the early 20th century that were devoted to the fostering of space exploration as a fundamental aspect of human destiny.

—. *Rockets into Space.* Cambridge, MA: Harvard University Press, 1990. This work by the rocket curator at the National Air and Space Museum of the Smithsonian Institution is perhaps the most useful and up-to-date short synthesis of the development of the rocket available in English.

Wolfe, Tom. *The Right Stuff.* New York: Farrar, Straus and Giroux, 1979. Wolfe gives an outstanding journalistic account of the first years of spaceflight, essentially Project Mercury, focusing on the Mercury Seven.

Zimmerman, Robert. *Genesis: The Story of Apollo 8.* New York: Four Walls Eight Windows, 1998. This work is a fine history of the most significant mission to the Moon other than Apollo 11.

Key Historical Analog Studies

Ashley, K. D. *Modeling Legal Argument: Reasoning with Cases and Hypotheticals.* Cambridge, MA: MIT Press, 1990. The author provides an application of analogs to the development of legal arguments.

Bartha, Paul F. A. *By Parallel Reasoning: The Construction and Evaluation of Analogical Arguments.* New York: Oxford University Press, 2010. This accessible study outlines the role of analogs in scientific reasoning.

Gentner, D., K. J. Holyoak, and B. N. Kokinov, editors. *The Analogical Mind: Perspectives from Cognitive Science.* Cambridge, MA: MIT Press., 2001. An important collection of essays applies analog theory to a range of topics in science, philosophy, law, sociology, and economics.

Hesse, Mary B. *Models and Analogies in Science.* Notre Dame, IN: University of Notre Dame Press, 1966. This enormously influential study shows why analogies are integral to understanding scientific practice.

Kyburg, H., and C. M. Teng. *Uncertain Inference.* Cambridge, U.K.: Cambridge University Press, 2001. Kyburg and Teng's work demonstrates the possibilities, as well as the pitfalls, of analog usage.

Mazlish, Bruce, editor. *The Railroad and the Space Program: An Exploration in Historical Analogy.* Cambridge, MA: MIT Press, 1965. NASA sponsored this study, an early effort to use analogs, to see what connections could be drawn between fostering the railroads and fostering space exploration.

Mitchell, M. *Analogy-Making as Perception.* Cambridge. MA: MIT Press, 1993. This work deals with the centrality and the ubiquity of analogy in creative thought among scientists, artists, and writers and questions the broad use of analogs for understanding and knowledge generation.

Neustadt, Richard E., and Ernest R. May. *Thinking in Time: The Uses of History for Decision Makers.* New York: Free Press, 1986. To overcome the temptation to use history incorrectly, the authors put forward a specific process for decision-makers in crisis situations and use case studies to highlight successes and failures in analog usage.

Made in United States
Orlando, FL
04 November 2026

90203416R00070